I0896239

Review of the U.S. Navy's Exposure Standard for Manufactured Vitreous Fibers

Subcommittee on Manufactured Vitreous Fibers
Committee on Toxicology
Board on Environmental Studies and Toxicology
Commission on Life Sciences
National Research Council

NATIONAL ACADEMY PRESS
Washington D.C.

NATIONAL ACADEMY PRESS 2101 Constitution Ave., N.W. Washington, D.C. 20418

NOTICE: The project that is the subject of this report was approved by the Governing Board of the National Research Council, whose members are drawn from the councils of the National Academy of Sciences, the National Academy of Engineering, and the Institute of Medicine. The members of the committee responsible for the report were chosen for their special competences and with regard for appropriate balance.

This project was supported by Contracts DAMD17-89-C-9086 and DAMD17-99-C-9049 between the National Academy of Sciences and the U.S. Department of Defense. Any opinions, findings, conclusions, or recommendations expressed in this publication are those of the author(s) and do not necessarily reflect the view of the organizations or agencies that provided support for this project.

International Standard Book Number 0-309-07093-7
Additional copies of this report are available from:
National Academy Press 2101 Constitution Ave., NW Box 285 Washington, DC 20055
800-624-6242 202-334-3313 (in the Washington metropolitan area) http://www.nap.edu

Printed in the United States of America

THE NATIONAL ACADEMIES

National Academy of Sciences
National Academy of Engineering
Institute of Medicine
National Research Council

The **National Academy of Sciences** is a private, nonprofit, self-perpetuating society of distinguished scholars engaged in scientific and engineering research, dedicated to the furtherance of science and technology and to their use for the general welfare. Upon the authority of the charter granted to it by the Congress in 1863, the Academy has a mandate that requires it to advise the federal government on scientific and technical matters. Dr. Bruce M. Alberts is president of the National Academy of Sciences.

The **National Academy of Engineering** was established in 1964, under the charter of the National Academy of Sciences, as a parallel organization of outstanding engineers. It is autonomous in its administration and in the selection of its members, sharing with the National Academy of Sciences the responsibility for advising the federal government. The National Academy of Engineering also sponsors engineering programs aimed at meeting national needs, encourages education and research, and recognizes the superior achievements of engineers. Dr. William A. Wulf is president of the National Academy of Engineering.

The **Institute of Medicine** was established in 1970 by the National Academy of Sciences to secure the services of eminent members of appropriate professions in the examination of policy matters pertaining to the health of the public. The Institute acts under the responsibility given to the National Academy of Sciences by its congressional charter to be an adviser to the federal government and, upon its own initiative, to identify issues of medical care, research, and education. Dr. Kenneth I. Shine is president of the Institute of Medicine.

The **National Research Council** was organized by the National Academy of Sciences in 1916 to associate the broad community of science and technology with the Academy's purposes of furthering knowledge and advising the federal government. Functioning in accordance with general policies determined by the Academy, the Council has become the principal operating agency of both the National Academy of Sciences and the National Academy of Engineering in providing services to the government, the public, and the scientific and engineering communities. The Council is administered jointly by both Academies and the Institute of Medicine. Dr. Bruce M. Alberts and Dr. William A. Wulf are chairman and vice chairman, respectively, of the National Research Council.

www.national-academies.org

SUBCOMMITTEE ON MANUFACTURED VITREOUS FIBERS

MORTON LIPPMANN *(Chair),* New York University School of Medicine, Tuxedo, New York
LEONARD CHIAZZE, Georgetown University School of Medicine, Washington, D.C.
DAVID B. COULTAS, University of New Mexico, Albuquerque, New Mexico
KEVIN E. DRISCOLL, Procter & Gamble Pharmaceuticals, Cincinnati, Ohio
AGNES B. KANE, Brown University, Providence, Rhode Island
JAMES E. LOCKEY, University of Cincinnati, Cincinnati, Ohio
ERNEST E. MCCONNELL, ToxPath, Inc., Raleigh, North Carolina
GÜNTER OBERDÖRSTER, University of Rochester Medical School, Rochester, New York
LORENZ R. RHOMBERG, Gradient Corporation, Cambridge, Massachusetts
MARK UTELL, University of Rochester Medical Center, Rochester, New York
DAVID B.WARHEIT, DuPont Haskell Laboratory, Newark, Delaware

STAFF

KULBIR S. BAKSHI, Program Director, Committee on Toxicology
ROBERTA M. WEDGE, Project Director
EILEEN N. ABT, Research Associate
NORMAN GROSSBLATT, Editor
MIRSADA KARALIC-LONCAREVIC, Information Specialist
LUCY V. FUSCO, Project Assistant
LINDA LEONARD, Project Assistant

SPONSOR

U.S. DEPARTMENT OF DEFENSE

COMMITTEE ON TOXICOLOGY

BAILUS WALKER, JR., *(CHAIR),* Howard University Medical Center and American Public Health Association, Washington, D.C.

MELVIN E. ANDERSEN, Colorado State University, Denver, Colorado

GERMAINE M. BUCK, University at Buffalo, State of New York

ROBERT E. FORSTER II, University of Pennsylvania, Philadelphia, Pennsylvania

PAUL M.D. FOSTER, Chemical Industry Institute of Toxicology, Research Triangle Park, North Carolina

WILLIAM E. HALPERIN, National Institute for Occupational Safety and Health, Cincinnati, Ohio

CHARLES H. HOBBS, Lovelace Respiratory Research Institute and Lovelace Biomedical and Environmental Research Institute, Albuquerque, New Mexico

SAMUEL KACEW, University of Ottawa, Ontario, Canada

NANCY KERKVLIET, Oregon State University, Corvallis, Oregon

FLORENCE K. KINOSHITA, Hercules Incorporated, Wilmington, Delaware

MICHAEL J. KOSNETT, University of Colorado Health Sciences Center, Denver, Colorado

MORTON LIPPMANN, New York University School of Medicine, Tuxedo, New York

ERNEST E. MCCONNELL, ToxPath, Inc., Raleigh, North Carolina

THOMAS E. MCKONE, Lawrence Berkeley National Laboratory and University of California, Berkeley, California

HARIHARA MEHENDALE, The University of Louisiana of Louisiana at Monroe, Louisiana

DAVID H. MOORE, Battelle Memorial Institute, Bel Air, Maryland

GÜNTER OBERDÖRSTER, University of Rochester, Rochester, New York

JOHN L. O'DONOGHUE, Eastman Kodak Company, Rochester, New York

GEORGE M. RUSCH, AlliedSignal, Inc., Morristown, New Jersey

MARY E. VORE, University of Kentucky, Lexington, Kentucky

ANNETTA P. WATSON, Oak Ridge National Laboratory, Oak Ridge, Tennessee

LAUREN ZEISE, California Environmental Protection Agency, Oakland, California

STAFF

KULBIR S. BAKSHI, Program Director

SUSAN N.J. PANG, Program Officer

ABIGAIL E. STACK, Program Officer

MICHELLE C. CATLIN, Research Associate

RUTH E. CROSSGROVE, Publications Manager

KATHRINE J. IVERSON, Manager, Toxicology Information Center

EMILY SMAIL, Project Assistant

BOARD ON ENVIRONMENTAL STUDIES AND TOXICOLOGY

COMMISSION ON LIFE SCIENCES

OTHER REPORTS OF THE BOARD ON ENVIRONMENTAL STUDIES AND TOXICOLOGY

Strengthening Science at the U.S. Environmental Protection Agency: Research Management and Peer Review Practice (2000)
Scientific Frontiers in Developmental Toxicology and Risk Assessment (2000)
Modeling Mobile-Source Emissions (2000)
Copper in Drinking Water (2000)
Ecological Indicators for the Nation (2000)
Waste Incineration and Public Health (1999)
Hormonally Active Agents in the Environment (1999)
Research Priorities for Airborne Particulate Matter: I. Immediate Priorities and a Long-Range Research Portfolio (1998); II. Evaluating Research Progress and Updating the Portfolio (1999)
Ozone-Forming Potential of Reformulated Gasoline (1999)
Risk-Based Waste Classification in California (1999)
Arsenic in Drinking Water (1999)
Brucellosis in the Greater Yellowstone Area (1998)
The National Research Council's Committee on Toxicology: The First 50 Years (1997)
Toxicologic Assessment of the Army's Zinc Cadmium Sulfide Dispersion Tests (1997)
Carcinogens and Anticarcinogens in the Human Diet (1996)
Upstream: Salmon and Society in the Pacific Northwest (1996)
Science and the Endangered Species Act (1995)
Wetlands: Characteristics and Boundaries (1995)
Biologic Markers (5 reports, 1989-1995)
Review of EPA's Environmental Monitoring and Assessment Program (3 reports, 1994-1995)
Science and Judgment in Risk Assessment (1994)
Ranking Hazardous Waste Sites for Remedial Action (1994)
Pesticides in the Diets of Infants and Children (1993)
Issues in Risk Assessment (1993)
Setting Priorities for Land Conservation (1993)
Protecting Visibility in National Parks and Wilderness Areas (1993)
Dolphins and the Tuna Industry (1992)
Hazardous Materials on the Public Lands (1992)
Science and the National Parks (1992)
Animals as Sentinels of Environmental Health Hazards (1991)

Assessment of the U.S. Outer Continental Shelf Environmental Studies Program, Volumes I-IV (1991-1993)
Human Exposure Assessment for Airborne Pollutants (1991)
Monitoring Human Tissues for Toxic Substances (1991)
Rethinking the Ozone Problem in Urban and Regional Air Pollution (1991)
Decline of the Sea Turtles (1990)

Copies of these reports may be ordered from
the National Academy Press
(800) 624-6242
(202) 334-3313
www.nap.edu

PREFACE

HE U.S. Navy Environmental Health Center (NEHC), part of the Bureau of Medicine and Surgery, oversees the Navy's Occupational Safety and Health Program. In 1995, the NEHC established an occupational exposure standard of 2 fibers per cubic centimeter for manufactured vitreous fibers to protect workers against adverse health effects. Recently, this standard was lowered to 1 fiber per cubic centimeter to comply with existing guidelines developed by other industrial hygiene organizations. In setting the earlier exposure standard, the Navy reviewed the toxicological and epidemiological studies available in the published scientific literature and the rationales used by the National Institute for Occupational Safety and Health in developing its recommended exposure limit of 3 fibers per cubic centimeter and the Occupational Safety and Health Administration in proposing a permissible exposure limit of 1 fiber per cubic centimeter. The Navy chose an occupational exposure limit that was an average of those two values.

In this report, the Subcommittee on Manufactured Vitreous Fibers of the National Research Council (NRC) Committee on Toxicology reviews independently the scientific validity of the Navy's exposure limit and determines whether any additional scientific studies should be considered by the Navy in choosing its exposure limit. To prepare the report, the subcommittee reviewed the materials supplied by the Navy, and by other organization's and individuals, and information gathered at a public meeting held at the J. Erik Jonsson Woods Hole Center, Massachusetts, on July 16, 1998. This report is intended to assist the

Navy in developing a process for establishing occupational exposure limits for other materials and to highlight concerns that might influence the choice of a protective value.

The subcommittee wishes to thank David A. Macys, program officer, of the Office of Naval Research, and Patricia Krevonick, senior industrial hygienist, of the NEHC, for their presentations at the public meeting and for their responses to written questions from the subcommittee. We also gratefully acknowledge John Hadley, corporate toxicologist of Owens Corning for making a presentation to the subcommittee.

This report has been reviewed in draft form by individuals chosen for their diverse perspectives and technical expertise, in accordance with procedures for reviewing NRC reports approved by the NRC's Report Review Committee. The purpose of this independent review is to provide candid and critical comments that will assist NRC in making the published report as sound as possible and to ensure that the report meets institutional standards for objectivity, evidence, and responsiveness to the study charge. The content of the final report is the responsibility of NRC and the study subcommittee, and not of the reviewers. The review comments and draft manuscript remain confidential to protect the integrity of the deliberative process. We wish to thank the following individuals, who are neither officials nor employees of NRC, for their participation in the review of this report: Andrew Churg, University of British Columbia; Walter Eastes, Owens Corning Corporation; Jeffrey Everitt, Chemical Industry Institute of Toxicology; Thomas Hesterberg, Johns Manville Corporation; Daniel Luchtel, University of Washington; Gary Marsh, University of Pittsburgh; Roger McClellan, Chemical Industry Institute of Toxicology (retired); and Vanessa Vu, U.S. Environmental Protection Agency. These reviewers have provided many constructive comments and suggestions; it must be emphasized, however, that responsibility for the final content of this report rests entirely with the authoring subcommittee and NRC.

I am also grateful for the assistance of the NRC staff in the preparation of this report. In particular, the subcommittee wishes to acknowledge Kulbir Bakshi, program director of the Committee on Toxicology; Roberta Wedge, staff officer for the subcommittee; and Eileen Abt, research associate. Other staff members who contributed to this effort are Norman Grossblatt, editor; Lucy Fusco and Linda Leonard, project assistants, and Mirsada Karalic-Loncarevic, information specialist.

Finally, I would like to thank the members of the subcommittee for their valuable expertise and dedicated efforts throughout the preparation of this report. Their efforts in preparing this report within a very short time are much appreciated.

Morton Lippmann, Ph.D.
Chair, Subcommittee on Manufactured Vitreous Fibers

CONTENTS

REVIEW OF THE U.S. NAVY'S EXPOSURE STANDARD FOR MANUFACTURED VITREOUS FIBERS

SUMMARY

ANUFACTURED vitreous fibers (MVF), also known as synthetic vitreous fibers, are generally considered to be less hazardous than asbestos to human health. They are used in a variety of thermal- and acoustical-insulation applications, frequently as an asbestos substitute or as a filtration medium. The Navy uses MVF in a number of shipboard and onshore applications.

To protect Navy personnel from potentially harmful exposures to MVF, the U.S. Navy Environmental Health Center (NEHC) developed occupational exposure standards. They are contained in the Navy's *Occupational Safety and Health Program Manual* and are supported by documentation in NEHC technical manuals. The documentation for the MVF occupational exposure standards assists industrial hygienists, occupational medicine physicians, and other Navy health professionals in assessing and controlling the health hazards associated with exposure to these fibers. Occupational exposures might result from such activities as construction, maintenance, operation, and decommissioning of naval ships and facilities. The exposure standards cover civilian and military personnel, but they do not apply to Navy contractors, who are regulated by the Occupational Safety and Health Administration (OSHA) and applicable state regulatory agencies.

In 1997, the National Research Council (NRC) was asked to conduct an independent review of the Navy's toxicological assessment of MVF and to evaluate the scientific validity of its exposure standard of 2 fibers per cubic centimeter of air (f/cm^3). The NRC assigned the task to the Committee on Toxicology, which convened the Subcommittee on Manufactured Vitreous Fibers, a multidisciplinary group of experts, to determine whether all relevant toxicological and epidemiological data were appropriately considered in developing the exposure standard; and to examine the uncertainty, variability, and quality of data and the appropriateness of assumptions used in the derivation of the exposure standard. The subcommittee was also asked to identify deficiencies in the MVF database and, where appropriate, to make recommendations for future research and data development.

THE SUBCOMMITTEE'S APPROACH

The Navy provided the subcommittee with its document *Man-Made Vitreous Fibers,* which is part of NEHC's Technical Manual (NEHC-TM6290.91-1 Rev. A) (NEHC 1997b), and its associated *Health Hazard Information Summary: Man-Made Vitreous Fibers* (NEHC 1997a). A background document, *Navy Exposure Limit for Man-Made Vitreous Fibers: A Retrospective Look at the Decision History* (Krevonick 1998), was also submitted to the subcommittee. During a public meeting, the subcommittee heard presentations on those documents from the Navy and a presentation on the toxicology of MVF from a manufacturer.

The subcommittee reviewed the Navy's documentation supporting the 2-f/cm^3 standard. The documentation addressed an array of topics related to MVF, including production and use, chemical and physical properties, sampling and analysis, standards and recommendations, exposure, and toxicological and epidemiological data. The subcommittee began its review with an overview of the manufacturing processes, chemical composition, and classification of MVF, including newly developed fibers. Changes in fiber chemistry resulting from use and thermal stress were also discussed. The subcommittee reviewed the Navy's supporting documentation regarding sampling techniques, analytical methods, and toxicological and epidemiological studies. In the case of several of those topics, the subcommittee identified data gaps or issues that needed to be more thoroughly addressed by the Navy, including dosimetry and fiber biopersistence. The process used by the Navy to derive its occupational exposure standard was also discussed.

In January 1999, the Navy revised its *Occupational Safety and Health Program Manual* (CNO 1999), changing the occupational exposure limit for MVF to the American Conference of Governmental Industrial Hygienists (ACGIH) threshold limit value (TLV) of 1 f/cm^3. The subcommittee subsequently expanded its review of the Navy's occupational exposure standard to include the new value. Finally, the subcommittee made recommendations on future research needs.

THE SUBCOMMITTEE'S EVALUATION

In general, the subcommittee found that the Navy made a good start in assessing the health effects of MVF. Further work is necessary, however, to ensure that military and civilian personnel are adequately protected when exposed to these fibers. Recent life-cycle-analysis studies for refractory ceramic fibers (RCF) have generated substantial knowledge on exposures of workers engaged in the manufacture, installation, and removal of RCF, but the data on after-manufacture exposures to other types of MVF are considerably less robust. Most of the monitoring data and epidemiological studies are based on workers in the MVF manufacturing sector. Although exposure data related to the installation and removal of RCF are available in the open literature, such data are generally lacking for other types of MVF in the published literature and in the Navy's documentation. The Navy provided little monitoring information on the types of fibers and concentrations associated with naval operations.

Manufacturing Processes, Chemical Composition, and Classification

The subcommittee determined that the Navy had described the fiber types and their chemical composition appropriately. The three fiber types discussed by the Navy are fibrous glass, rock and slag wools, and RCF, all of which have silica backbones but are produced with different technologies from different materials. The subcommittee noted, however, that the Navy included very little information on the physical or chemical degradation of MVF or on the effects of time and stress on their composition. The subcommittee recommends that the Navy be alert to the potential risks posed by degraded fibers and by new types of fibers.

Sampling, Analytical Methods, and Exposure Assessment

In reviewing the Navy's sampling procedures and analytical methods for determining airborne fiber concentrations, the subcommittee concurred with the Navy's decision to use a standard based on fiber count rather than one based on gravimetric methods. In reviewing the Navy's exposure assessment, the subcommittee concluded that the Navy's documentation should provide specific guidance on fiber-counting methods that are considered most appropriate for monitoring particular fiber lengths and diameters. Furthermore, a more up-to-date review of the effects of fiber morphology on deposition in and clearance from the lung would be helpful to assist the Navy in determining which MVF are likely to pose greater risks to workers and therefore require more stringent exposure limits.

The Navy's documentation included a review of environmental exposures and numerous studies on worker exposure; however, most of the latter studies focused on the manufacture of MVF. The subcommittee did not consider the studies reviewed by the Navy to be particularly relevant for determining whether the Navy's occupational exposure standard would be protective of military and civilian personnel. Although the Navy apparently monitors MVF in the workplace, it made no attempt in its documentation to link measured air concentrations to exposures of workers who install, use, maintain, or remove MVF.

Biopersistence of Vitreous Fibers

The ability of vitreous or asbestos fibers to cause disease is based on biopersistence (the time that an intact fiber remains in the lung), which in turn depends on the fiber's chemical composition and physical dimensions. The subcommittee concluded that the Navy's documentation does not provide an adequate assessment of the role of fiber biopersistence in health effects. To address that issue, the subcommittee presents several studies that have been conducted on fiber solubility and biopersistence in the lung. Longer fibers, in general, are more biopersistent, more toxic to cells, and more mutagenic.

Toxicological Studies

Although the Navy reviewed many of the acute and chronic toxicity studies conducted on animals, the subcommittee concluded that the reviews were not sufficiently critical of the animal models or the methods used to dose the animals. For instance, the Navy did not comment on the relative merits of the cited studies. The subcommittee notes that inhalation studies provide better toxicological information because they qualitatively (although not quantitatively) most closely resemble human exposures. Studies that use intratracheal instillation might be valuable for a preliminary assessment of a fiber's toxicity, but they should not be used for establishing exposure standards. Similarly, intracavitary dosing studies should be used with caution because they might yield false-positive results on the development of lung cancer and mesothelioma.

Epidemiological Studies

The Navy was thorough in its summary of epidemiological studies of workers manufacturing MVF, focusing primarily on three large cohort studies in North America and Europe. However, the Navy did not cite several recent studies that re-examined those cohorts or studies that identified new cohorts, such as populations of workers manufacturing RCF. Such omissions are important because recent animal studies suggest that RCF and other very biopersistent fibers might have carcinogenic potential and therefore might warrant special consideration when an occupational exposure standard for MVF is developed. In addition, the Navy did not present a critical assessment of the limitations of the epidemiological studies. For instance, potential confounding factors, such as smoking, can affect the conclusions that can be drawn from the large studies of workers who manufacture MVF.

Evaluation of the Navy's Exposure Standard

The subcommittee reviewed the process and rationale used by the Navy to adopt the occupational exposure standard of 2 f/cm^3 for MVF in 1995. That standard was higher than the 1992 proposed OSHA permissible exposure limit (PEL) of 1 f/cm^3, but it was lower than the 1977 National Institute for Occupational Safety and Health (NIOSH) recommended exposure limit (REL) of 3 f/cm^3. The Navy acknowledged that it had no scientific justification for selecting the value of 2 f/cm^3 instead of the NIOSH or OSHA limits and that the choice was essentially the average of the two values. However, given the Navy's stated goal of having a reasonable yet protective fiber standard, the subcommittee concludes that the adoption of the OSHA PEL of 1 f/cm^3 would have been more conservative. Alternatively, the Navy could have derived its own occupational exposure standard by conducting a risk assessment based on a rigorous scientific review of existing toxicological and epidemiological data.

In January 1999, the Navy policy for establishing occupational exposure standards was changed. That resulted in the adoption of the ACGIH TLV of 1 f/cm^3. The subcommittee believes that lowering the standard from 2 to 1 f/cm^3 is appropriate on the basis of available toxicological and epidemiological data and is in accord with other national and international occupational health limits for MVF.

The subcommittee concludes that 1 f/cm^3 might not be protective for RCF or for some other MVF that are particularly biopersistent. The subcommittee recommends that the Navy consider setting separate exposure standards for the more biopersistent fibers.

Information Gaps and Recommendations for Future Research

Future research efforts should focus on identifying and monitoring the health effects associated with different fiber types. Animal models that are most appropriate for human health risk assessment of MVF should be identified. The development of short-term screening assays to predict long-term effects should encourage the testing of new fibers for health effects. Epidemiological studies should look not only at workers engaged in the manufacture of MVF, but also at those involved in their installation, maintenance, and removal. Data on the latter types of exposure are available for RCF but are less robust for other MVF, especially with regard to exposures during MVF removal. Monitoring studies of workers exposed to new, used, or stressed fibers would add considerably to the understanding of the health effects of both short-term and long-term exposures to these fibers.

1

INTRODUCTION

BACKGROUND

ANUFACTURED vitreous fibers (MVF) are fibrous, inorganic materials derived from various minerals. They are also referred to as man-made vitreous fibers, man-made mineral fibers, and synthetic vitreous fibers. MVF are manufactured by melting the raw materials and then forming fibers by several methods, including a drawing process that produces continuous fibers that are used in textiles; a rotary spraying process that produces fibers for insulation; a flame attenuation process that produces specialty fibers such as those used for filtration; a blowing process that is typically used to produce refractory ceramic fibers (RCF); and a wheel-centrifuge process used for RCF, rock wool, and slag wool insulation fibers (TIMA 1993). MVF are designed and developed to meet specific requirements, such as providing insulation at various temperatures, resisting degradation during filtering, and sound proofing.

Historically, MVF have been grouped into classes on the basis of the primary materials from which they are made. For instance, glass fibers are made from sand, rock fibers from basalt, slag fibers from smelter residues, and RCF from clay. However, with the introduction of new fibers and new production processes over the last decade, that classification system is no longer adequate. Changes in the chemistry of fibers in all the classes noted above have resulted in overlap of various classes and fibers that no longer appropriately fit into any of the classes. In addition, the types of fibers in a given class can be variable; for instance, glass fibers vary widely in biologic potential (McConnell et al. 1999) as a result of differences in their chemistries and production methods (ACGIH 1997).

Little consideration was given to potential health effects of exposures to MVF until the hazards posed by inhaled asbestos—including asbestosis, lung cancer, and mesothelioma—were identified in the early 1960s (Wagner et al. 1960; Bader et al., 1961). Asbestosis was the first disease that clearly defined the potential for inhaled fibers to cause human health effects. The first case of asbestosis was described in 1907 by Murray, who reported that a carding machine operator died of pulmonary injury associated with diffuse pulmonary fibrosis. In 1924, Merewether and Price (1930), reporting on a study of British workers noted the unequivocal relationship between asbestos and pulmonary fibrosis. In the 1950's and 1960's, the association between asbestos exposure and the development of lung cancer (Doll 1955) or mesothelioma (Wagner et al. 1960) was first recognized. Only recently has the latent nature of asbestos-related disease been fully appreciated, particularly in the case of mesothelioma, which can occur more than 20 years after exposure. Gilson (1966) concluded that the average latent interval for the development of carcinoma of the lung was 20 years, whereas the interval between first exposure to asbestos and onset of symptoms of mesothelioma can be 25-50 years, with an average latency of 33 years. During World Wars I and II, there was a need to insulate ships rapidly; as a result, the number of asbestos-exposed individuals increased. Their exposure was correlated with an increased incidence of asbestos-related lung and pleural disease, often manifesting itself 1 or 2 decades after the end of exposure (Kennedy and Kelly 1993).

MVF were designed as replacements for asbestos in various applications, including insulation. In light of the asbestos legacy and the absence of a toxicological database on vitreous fibers combined with the latency issue, it is not surprising that concerns have been raised about human health effects related to MVF exposure. Because of the many types of fibers in use in the 1970's, including MVF, a series of studies were conducted to evaluate their pathologic potential. Numerous types of fibers were instilled into the animal's pleural cavities (Stanton and Wrench 1972) or injected into animal's abdominal cavities (Pott and Friedrichs 1972) primarily to evaluate their carcinogenicity. Both groups of investigators found that the dimensions of the fibers were critical to their pathogenicity. Stanton et al. (1981) proposed that although fibers greater than 8 μm in length and less than or equal to 0.25 μm in diameter might be more carcinogenic, the ratio of fiber length to width—the aspect ratio—was more important for carcinogenicity than either dimension alone. Those studies were the original basis of two of the three legs of the "3-D" concept—dose, dimension, and durability. The importance of durability quickly followed when it was recognized that a fiber had to reside in the lung for a relatively long period before it could cause chronic disease.

Until 1984, the Navy based its occupational standard for exposure to fibrous glass dust and vitreous fibers on a gravimetric standard chosen with reference to the then operative Threshold Limit Value (TLV®) of the American Conference of Governmental Industrial Hygienists (ACGIH) for such dusts and on the permissible exposure limits of the Occupational Safety and Health Administration (OSHA). The guidelines (ACGIH) and standards (OSHA) treated glass fibers as nuisance dusts that were not otherwise regulated. The standard applied only to the mass of respirable particles, not to the particular toxicological properties of the material. The adequacy of this approach was questioned when some durable fibers became suspected of being animal carcinogens and when the dimensional characteristics of asbestos fibers were implicated in cancer in shipyard workers chronically exposed to asbestos. Those developments suggested that standards would be more soundly based on the number of WHO (World Health Organization)[1] fibers, especially those greater than 20 μm in length, rather than on their mass and that differentiation among fiber types based on their toxicity might be warranted.

Between 1984 and 1985, the Navy adopted an exposure standard of 2 fibers per cubic centimeter of air (f/cm^3) for all MVF. The standard was based largely on the practical difficulty of distinguishing among fiber types when assessing exposures in workplace operations and on the prevailing standard of 2 f/cm^3 for asbestos fibers, on the grounds that treating all fibers as though they were as toxic as asbestos would ensure sufficient protection. That was a conservative position, because asbestos fibers were thought to be substantially more toxic than most vitreous fibers in light of responses in animals to intratracheal instillations and intraperitoneal implantations in animals.

That approach became increasingly impractical as the OSHA standard for asbestos was lowered from 2 to 1 f/cm^3 and then to 0.1 f/cm^3. Because of the questionable relevance of this standard to vitreous fibers and the increasing burden posed by adhering to an asbestos-based

[1] WHO fibers are defined as having a diameter less than 3 μm, a length greater than 5 μm, and an aspect (length/width) ratio greater than 3:1 (WHO 1985).

standard, the Navy sought to establish a standard specific to MVF and adopted a standard of 2 f/cm^3 in 1995. It was based on the Navy's review and evaluation of published research, on manufacturers' recommendations, and on practical aspects of implementation.

In 1997, ACGIH proposed to revise the TLV for synthetic vitreous fibers from 2 f/cm^3 to 1 f/cm^3 (ACGIH, 1997); in 1998, it adopted the 1-f/cm^3 limit. In a laudable desire to keep the exposure of Navy personnel in line with health-protection standards established in other sectors, the Navy adopted a policy of complying with the 1-f/cm^3 TLV although this value was never formally adopted as the Navy's exposure standard. The Navy noted, however, that the exposure standards and recommendations of ACGIH, the National Institute for Occupational Safety and Health, and other institutions around the world are not strongly supported by quantitative evidence of potential risks, whether associated with cancer or chronic noncancer effects. There is a substantial amount of mortality data on workers involved in glass fiber—as well as rock and slag wool—manufacturing in the United States and Europe. Mortality data on RCF manufacturing are being collected in the United States and Europe, but the number of cases are insufficient for any definitive analyses. Some morbidity data on mineral wool and glass production workers do exist and are more comprehensive than morbidity data on RCF production workers. There are very few morbidity and mortality data on end users of MVF. Collectively, the data show little evidence of positive effects, and the consistency of the results and their applicability to the evaluation of health-protective inhalation standards in humans are uncertain. Consequently, although it is desirable to base exposure standards for MVF on toxicological and epidemiological data specific to the fibers in question, and quantitative dose-response analyses are preferred for providing confidence that numerical standards are set appropriately, existing standards are not supported by clearly articulated arguments for a causal association based on these data.

Several questions arise from those observations. Even if the Navy established a standard chosen to reflect exposure levels established by other organizations, such as the ACGIH 1-f/cm^3 guideline, what assurance is there that health protection would be achieved? Is a standard set in this way substantially stricter than necessary? For instance, could the Navy's standard of 2 f/cm^3 be justified as sufficient to protect health? Is there evidence that different standards might be necessary for different categories of MVF?

THE SUBCOMMITTEE'S TASK

To address the above questions, the Navy Environmental Health Center (NEHC) reviewed the available information on the relevant properties of fibrous glass, rock and slag wools, and RCF, in an October 1997 document, *Man-Made Vitreous Fibers* (NEHC-TM 6290.91-1 RevA); in its associated *Health Hazard Information Summary* (NEHC 1997a); and in a further document, *Navy Exposure Limit for Man-Made Vitreous Fibers: A Retrospective Look at the Decision History* (Krevonick 1998). The Navy requested that the National Research Council (NRC) review those documents to determine whether the Navy's recommended exposure limit of 2 f/cm^3 is scientifically valid. NRC assigned the project to the Committee on Toxicology, which convened the Subcommittee on Manufactured Vitreous Fibers. The subcommittee was assigned the following task:

> A study will be conducted to review the Navy's toxicological assessment of manufactured vitreous fibers (MVF) and evaluate the scientific validity of the Navy's recommended exposure limit of 2 fibers/cubic centimeter of air. The subcommittee will determine whether all relevant toxicity and epidemiology data were appropriately considered in developing the exposure limit. The uncertainty, variability, and quality of data and the appropriateness of the assumptions used in the derivation of the exposure limit will also be reviewed. Deficiencies in the database on MVF will be identified, and where appropriate, recommendations for future research and data development will be made.

During its review of the Navy documents, it became apparent to the subcommittee that potential exposure of Navy personnel to MVF were unlikely to be the same as that experienced by workers involved in the manufacture of MVF. The subcommittee subsequently asked the Navy to provide whatever exposure information it had on the nature of exposures of Navy personnel. During the course of the subcommittee's review of the Navy's 1997 documentation, the Navy changed the process by which it adopts an occupational exposure limit. After the Navy had given the subcommittee the requested exposure information, it requested that the subcommittee comment, if possible, upon the new occupational exposure limit of 1 f/cm^3 and the process used by the Navy to select it.

ORGANIZATION OF THE REPORT

In the remainder of this report, the subcommittee reviews the Navy's documentation and provides an assessment of the Navy's approach for setting occupational exposure limits. Chapter 2, Chapter 3, Chapter 4, Chapter 5 through Chapter 6 provide commentary on the adequacy and completeness of the Navy's review of the relevant scientific literature and its interpretation. Specifically, Chapter 2 discusses information on manufacturing processes, chemical composition, and classification of fibers; Chapter 3 assesses data on sampling, analytical methods, and dosimetry; Chapter 4 reviews biopersistence of vitreous fibers; Chapter 5 discusses toxicological studies on MVF; and Chapter 6 addresses the epidemiological studies. Chapter 7 reviews the Navy's approach to selecting an exposure limit and offers recommendations on how the Navy could conduct the kind of scientific evaluation and risk analysis necessary to establish a protective exposure limit. Chapter 8 addresses information gaps and topics for future research.

2

MANUFACTURING PROCESSES, CHEMICAL COMPOSITION, AND CLASSIFICATION

HE Navy's report *Man-Made Vitreous Fibers,* discusses three principal classes of manufactured vitreous fibers (MVF): fibrous glass, mineral wool (slag and rock wool), and refractory ceramic fibers. The manufacturing processes and chemical composition of each fiber class varies depending on their specific end use; however, categorizing fibers into these classes does not capture the diversity of fibers in each class or the different potential hazards associated with each class.

The last several years have seen an increase in the use of MVF in a variety of insulation and other industrial applications, including use in the Navy, most often as a replacement for asbestos. The Navy documentation on MVF discusses their manufacture and use in only general terms. The Navy does not indicate any applications that are specific to its needs other than a reference to the use of fibers in thermoplastics reinforcement for aircraft and marine hulls (NEHC 1997a). Subsequent information on exposure monitoring conducted by the Navy, informally received from the Navy Environmental Health Center, indicated that the Navy uses fibrous glass and wool primarily as insulation and, for continuous fibrous glass, in advanced composite material applications as reinforcement (P. Krevonick, Navy Environmental Health Center, Personal Commun., October 7, 1999). The monitoring data suggest the presence of MVF in boat structures, piping, acoustic panels, and lagging, although this information was not included in the Navy's original documentation. The lack of information on the specific materials that contain MVF and the quantities used by the Navy is of concern to the subcommittee; it makes it difficult to determine the extent of potential exposure of Navy personnel. These exposure standards cover all naval civilian and military personnel but do not apply to Navy contractors, which are regulated by the Occupational Safety and Health Administration or applicable state regulatory agencies.

The Navy does present an overview of the chemical composition of the fiber classes. Table 2-1 shows the chemical composition of various classes of fibers, expressed as percentages. From Table 2-1, it is evident that all MVF contain silica (SiO_2), but they vary widely in their other components, both between and within classes of fibers. It should be noted that although the Navy refers briefly to the use of special purpose fibers, it does not include any information on their composition or sizes or on what distinguishes them from other MVF. The manufacturing processes for and composition of each fiber type are discussed below.

GLASS FIBERS

The primary ingredient in glass fibers is naturally occurring silicon dioxide; it also contains small amounts of other minerals. Permutations are made by adding other substances, such as oxides of aluminum, titanium, and zinc as stabilizers and oxides of magnesium, lithium, barium, calcium, sodium, and potassium as modifiers. By varying the amounts and types of stabilizers and modifiers, one can alter the physical properties of glass fibers. Stabilizers contribute to chemical durability; the intended use determines the amount of stabilizer added.

Glass fibers are produced by mixing and melting the raw materials in high temperature furnaces and then processing them with various methods that depend on the end product. A continuous filament process is used for textile fibers, a rotary spray process for glass wool, and a flame attenuation process for making special purpose glass fibers.

In the manufacture of textile fibers, molten glass is continuously drawn from the melting pot through bushings. This process allows for little variation in the preset average fiber diameter, which typically ranges from 3 to 25 μm. These continuous glass filaments are used in various applications, including textiles, and as reinforcements for plastic composites, such as boat hulls and automobile body parts.

Glass wool is manufactured with a rotary process that consists of pouring the molten glass through a spinner that fiberizes the glass into discontinuous fibers. Fiber diameters vary widely: some are as small as

1 μm and the average is 3-15 μm. The glass wool fibers are bound together with such agents as urea-phenolic resins, which undergo a heat curing process that converts the binders to insoluble polymers. Other agents, such as lubricants and antistatic and wetting agents, can be

TABLE 2-1 Typical Chemical Composition of Some Commercial MVF[a]

	Composition (%)									
Name[b]	11	A	C	21	F	G	22	RCF-1	X-607	Insofrax
Class	Glass	Glass	Glass	Rock	Rock	Rock	Slag	RCF	RCF sub.[c]	RCF sub.[c]
Components										
SiO_2	63.40	65.00	61.70	46.30	56.30	60.10	38.40	47.70	58.30	76.20
Fe_2O_3	0.30	0.10	0.10	13.20	0.30	6.10	0.00	1.00	0.10	0.30
TiO_2	0.06	0.02	0.02	2.60	0.10	0.05	0.50	2.10	0.05	0.08
Al_2O_3	3.90	1.90	1.00	13.50	3.20	0.40	10.60	48.00	1.30	1.40
CaO	7.40	7.40	7.20	10.00	26.10	18.80	38.00	0.07	38.70	0.20
MgO	2.80	2.60	2.90	9.10	6.40	8.30	9.90	0.08	0.40	21.50
Na_2O	15.40	16.10	16.10	3.10	3.20	5.50	0.40	0.00	0.30	0.07
K_2O	1.30	0.70	0.60	1.40	0.70	0.20	0.50	0.20	0.10	0.10
B_2O_2	4.50	4.70	9.20	0.00	0.00	0.00	0.00	0.01	0.00	0.00
P_2O_5	0.00	1.10	1.10	0.40	2.90	0.08	0.00	0.10	0.40	0.03
SO_3	0.30	0.03	0.20	0.00	0.00	0.05	1.80	0.00	0.00	0.00
Cr_2O_3	0.00	0.00	0.00	0.04	0.00	0.00	0.00	0.03	0.00	0.00
MnO	0.01	0.00	0.01	0.20	0.00	0.00	0.70	0.00	0.00	0.01
ZrO_2	0.00	0.00	0.00	0.00	0.00	0.00	0.00	0.10	0.00	0.00
Total	99.40	99.60	100.00	99.80	99.10	99.50	100.80	99.40	99.30	99.90

[a] Data derived from Bernstein et al. (1996), Maxim et al. (1999a), and McConnell et al. (1994, 1995, and 1999); Material Safety Data sheets for Isofrax fibers from the Unifrax Corporation, Niagara Falls, NY.

[b] Name: 11, Certain Teed B glass wool fiber; A, new glass wool; C, new glass wool; 21, rock wool; F, rock wool; G, rock wool; 22, slag wool; RCF-1 - kaolin-based refractory ceramic fiber; X-607, rock wool produced by Unifrax; Isofrax, refractory ceramic fiber.

[c] Substituted RCF.

added in the production process. Glass wool is used in industrial and commercial insulation applications—such as batts, blankets, and blowing wool—and for air ducts, ceiling panels, and acoustic panels.

Special purpose glass fibers are produced with a flame attenuation process. The hot, molten glass is poured in front of a high temperature gas flame; this results in fibers with a mean diameter of less than 3 μm. Various types of binders can be added in the process, depending on the intended end use applications. Special purpose glass fibers are typically used in applications that require high thermal and acoustic insulation, as in the aircraft industry, and for filtration media.

MINERAL WOOL

In the United States, most mineral wool (slag wool and rock wool) production uses slag as a raw material. Slag is formed during the reduction of iron ore to pig iron. Modern slag wool is composed of calcium, magnesium, and aluminum silicates with trace amounts of other oxides; raw materials—including clay, sand, and limestone—can be added to the coke-fired cupola or melted in an electric or gas heated furnace. Rock wool is produced through the same process with basaltic rock, limestone, clay, and feldspar, and small amounts of other additives. The production of slag wool and rock wool includes a wheel centrifuge process that results in discontinuous fibers averaging 3.5-7 μm in diameter. As with fibrous glass, the manufacturing process produces a range of fiber diameters, including respirable fibers. The addition of urea-phenolic resin produces bonded wool that is typically used for insulation batts, boards, blankets, and pipe covering. Nonbonded mineral wool is used as blown insulation or in the production of ceiling tiles.

REFRACTORY CERAMIC FIBERS

RCF comprise 1-2% of the worldwide production of MVF and are used in high temperature specialty applications. RCFs are used as bulk fibers, blankets, boards, paper, and textile products. They are produced by melting and spinning or blowing calcined kaolin or a mixture of alumina and oxides of zirconium, boron, or titanium. The average diameter of RCFs is 1-5 μm. RCF are unique in that although they are initially amorphous, they can be partially converted to a crystalline form such as mullite or cristobalite, when heated to above 1800°F.

CONCLUSIONS

Each of the three categories of fibers contains fibers with different chemical compositions and different sizes. Animal and human studies that address potential health consequences of exposures to MVF should be based on knowledge of the chemical and size characteristics of the fibers under study.

The Navy's review of the production, use, and chemical and physical properties of MVF covers technologies only through 1993. Of particular concern to the subcommittee is the Navy's understanding of the effects of time and temperature on the composition of MVF. The Navy does indicate in the Section "Chemical and Physical Properties" in *Man-Made Vitreous Fibers,* that MVF have high melting points, which make them good candidates for some applications, such as high temperature insulation, but it does not cite any studies on the wearing of these fibers and what happens to them when they are exposed to high temperatures. Since the anticipated exposure of Navy personnel is primarily to worn fibers, the subcommittee believes it would be helpful if the Navy included any relevant references on this topic or indicated that relevant data were not available.

Because of the dynamic nature of the development of "new" fibers, which are being used in a myriad of applications, one can expect that MVF in the future will be different from those in use or in production today. Therefore, the Navy will have to be cognizant of those differences both with regard to current and future use, but also, and just as importantly, with regard to "tear out" and replacement of older fibers. More recent advances in the production of MVF are not included in the Navy's documentation (Maxim et al. 1999b). New uses for MVF and the properties of the fibers may have a substantial impact on the types of exposures that may be anticipated for Navy personnel, now and in the future.

3

SAMPLING, ANALYTICAL METHODS, AND EXPOSURE ASSESSMENT

N this chapter, the subcommittee reviews the air sampling methodology recommended by the Navy for determining fiber concentrations in workplace air and the monitoring studies of ambient and workplace air summarized by the Navy in the "Exposure Data" section of its report, *Man-Made Vitreous Fibers.*

SAMPLING AND ANALYSIS

The discussion of air sampling and analysis provided in the Navy's review of manufactured vitreous fibers (MVF) is a straightforward description of standard procedures for industrial hygiene practice. It can be improved and updated with respect to gravimetric analysis and fiber counting.

1. Gravimetric Analysis

   When the American Conference of Governmental Industrial Hygienists (ACGIH) adopted a gravimetric limit of 5 mg/m^3 for continuous filament glass fibers in 1997, it specified that the sampler's inlet should accept the inhalable fraction (ACGIH 1997). Therefore, the National Institute for Occupational Safety and Health (NIOSH) analytical methods, 0500 and 0600 (NIOSH 1994), cited in the Navy's review, for otherwise unregulated total and respirable particles, respectively, are not directly applicable to MVF without modification.

Neither a total particle or respirable particle inlet sampler is appropriate for sampling the inhalable fraction of continuous filament glass.

The discussion of scanning electron microscopy (SEM) and transmission electron microscopy (TEM) in the section on gravimetric analysis appears to be misplaced, in that electron microscopy is more applicable to fiber counting than to mass measurement. A better example of an alternative means of determining sample mass would be x-ray fluorescence analysis based on the metal content of the MVF.

2. Fiber Counting

The introductory paragraphs on fiber counting are appropriate, but the purpose and intent of the remainder of the section are less clear. The cautionary notes about the limited resolution of phase-contrast optical microscopy (PCOM) and SEM for fine fibers (diameter less than 0.3 μm) and superfine fibers (diameter less than 0.1 μm) are appropriate. However, although guidance is given on how to describe the concentration of fibers resolvable with SEM and TEM, there is no guidance provided for determining fiber concentrations with PCOM. Should NIOSH B rules apply to SEM and TEM measurements? If not, why not? If so, should the analysis list separately the fibers seen with SEM or TEM that have diameters above and below about 0.3 μm? If that were done, the concentration of fibers with diameters greater than 0.3 μm might be comparable with the concentration obtained with PCOM, whereas fibers smaller than 0.3 μm might indicate the limitations of PCOM analysis for the particular exposure and risk assessment.

It might also be appropriate to note that using the resolving power of an electron microscope to count the number concentration of fibers less than 5 μm in length is inappropriate because the hazard is largely confined to fibers longer than 5 μm. While the 5-μm length limit has long been used and is incorporated in the NIOSH and ACGIH Threshold Limit Value (TLV) protocols, there is evidence that a better length limit may be 10 or 20 μm (HEI 1991; Lippmann 1994). In any case, an important need when scanning electron microscopic images of sampled fibers is to measure the length distribution of at least several hundred fibers longer than 5 μm. Furthermore, microscopic fields to be counted should be separated by 10 fields to ensure that the fiber distribution is valid (Miller 1999).

EXPOSURE ASSESSMENT

Understanding the risks to workers posed by exposure to MVF requires a knowledge not only of the toxicological effects of MVF and the dose-response relationships, but also of the environmental concentrations to which workers are likely to be exposed when performing maintenance or removal that disturbs the fibers or materials that contain them. The Navy document *Man-Made Vitreous Fibers* contains a short overview of environmental, nonoccupational exposure to MVFs and a discussion of occupational exposures, including ocular, dermal, as well as upper respiratory irritation, and inhalation hazards. The Navy document reviews MVF air sampling data by fiber type during manufacture, installation, use, and removal of MVF.

The subcommittee has two concerns regarding the Navy's review of air sampling data from fiber production operations. First, the Navy is not engaged in the manufacture of MVF, so these data collected during manufacturing appear to have little direct applicability to the Navy. Information on concentrations of MVF to which insulation installers or removers can be exposed is relevant to naval workers, but the Navy does not state whether it conducts monitoring of workers who could be exposed to MVF through direct contact, such as the installation or removal of MVF, or through indirect contact, such as proximity to equipment that contains MVF.

The Navy documentation of monitoring data on worker exposures to MVF during manufacturing and use is important for linking exposures to the results of epidemiological studies that examined adverse effects of such exposures. At the same time the Navy must clearly determine whether its proposed occupational exposure limits will be protective for its workers. To do that, the Navy should have presented the results of any exposure monitoring it conducted. The Navy provided the subcommittee with very limited information on its exposure monitoring during the preparation of this report (P. Krevonick, Navy Environmental Health Center, Personal Commun., October 7, 1999) but indicated that such monitoring had been conducted since January 1980. The monitoring information that was presented was cursory and difficult to understand and failed to indicate when the monitoring was conducted, how many workers were exposed, and other basic methodological and evaluative information. Since the Navy has been collecting at least some types of monitoring information since 1980 in its Navy Occupational Exposure

Database, the subcommittee does not understand why these data, or a summary of them, were not included in *Man-Made Vitreous Fibers* with some discussion of exposures, operations that result in high ambient fiber concentrations, and when current or proposed occupational exposure limits have been exceeded and by how much. Such information should have been presented in the Navy's original documentation with some correlation between air concentrations and exposures of naval personnel.

The Navy should also have discussed the representativeness of the monitoring information in terms of worker categories. For example, if the Navy industrial hygienists monitored fiber concentrations in air at one site where MVF-containing materials were being sanded, are these concentrations likely to be indicative of those at other sanding sites or sites of similar applications? If workers are engaged in sanding at other sites, are there likely to be differences in exposure related to configuration of the work environment or the extent of ventilation-related controls, with respect to the types, ages, thermal stress histories, the size of the objects being sanded, the abrasives being used, or the mechanical energy applied to the sander? Answers to questions like those would enable both the Navy and the subcommittee to determine which workers have the greatest exposure, whether the worst-case exposures exceed the occupational exposure limits, and, if so, what measures can be taken to reduce the exposures. Failure of the Navy to provide the monitoring information it has is a serious flaw in its documentation. At the least, the Navy document should have linked the studies it reviewed to what was known about exposures of Navy personnel.

CONCLUSIONS

With regard to air sampling for MVF, there needs to be clear and explicit guidance on the collection of representative and biologically relevant samples of particulate fibers longer than 5 μm for analyses of fiber number concentrations. In addition, installation and removal operations create high dust levels, and there is a need for guidance on the sampling of MVF dusts for analyses of mass concentrations of inhalable particulate matter as a nuisance dust.

NIOSH recommends PCOM as an appropriate analytical method for fiber counting of conventional glass fibers, rock and slag wools, and RCF with fiber diameters typically greater than 1 μm. However, that recom mendation should be modified to record separately the total number of fibers longer than 5, 10, and 20 μm. The subcommittee makes the same recommendation for determining concentrations of special-purpose MVF, which have a large fraction of fibers less than 1 μm in diameter.

For maintenance and removal operations where occupational exposures are likely to include high dust concentrations instead of or in addition to high fiber number concentrations, samplers with inlets that meet the ACGIH-ISO-CEN (American Conference of Governmental Industrial Hygienists-International Standards Organization-Comité Europeen de Normalisation) criteria for inhalable particulate matter should be used, and gravimetric determinations of the dust should be made.

4

BIOPERSISTENCE OF VITREOUS FIBERS

HE Navy addresses the biopersistence of vitreous fibers in the section of *Man-Made Vitreous Fibers* titled "Fiber Morphology, Deposition, and Clearance" which presents studies on the deposition and dissolution of various types of fibers. As discussed in detail later in this chapter, the subcommittee found the Navy's discussion of biopersistence, the primary determinant of fiber toxicity, to be inadequate with regard to its review of fiber morphology, deposition, and clearance. Furthermore, the Navy failed to cite many relevant publications, including some that provide important insights into the factors affecting target dose, biological responses to inhaled manufactured vitrous fibers (MVF), and methods for assessing biopersistence. To address some of those information gaps, the subcommittee here defines biopersistence, and how to assess and measure it.

DEFINING BIOPERSISTENCE

Biopersistence with regard to MVF refers to the length of time that an intact fiber remains in the lung and pleura (thorax). As first defined at a 1988 meeting of the International Programme on Chemical Safety (IPCS 1988), the term persistence refers to the ability of a fiber to stay in the biological environment; the prefix "bio" has been added to better delineate the definition. This definition is still applicable, although it has been modified to encompass the idea of translocation to target cells and tissues. The importance of this expanded definition is based on the understanding that the biopersistence of MVF in vulnerable tissues of the thorax is the primary determinant of its pathogenicity: MVF must remain in the tissues long enough for chronic disease to occur (Berry 1999).

Given the importance of understanding biopersistence when determining protective exposure levels, the subcommittee found the Navy's discussion of the following aspects of biopersistence to be inadequate: critical target region, fiber size in relation to penetration and deposition, and in vivo dissolution of fibers.

In the Navy's discussion of the critical target region, it appears to assume that fibers that do not reach the alveoli are of no concern. However, the subcommittee believes that it is important for the Navy to consider very large (noninhalable) fibers that might be of concern for skin irritation, and thoracic fibers that deposit in the lung conductive airways and theoretically might be responsible for bronchitis and bronchogenic cancer in exposed occupational cohorts (Lippmann 1990a,b, 1993).

With regard to the Navy's discussion of fiber size in relation to penetration and deposition in lung regions, the subcommittee notes that in adults breathing orally, fibers with aerodynamic diameters of about 10 μm (physical diameter, about 3.3 μm) will penetrate only into the thorax, whereas most fibers with aerodynamic diameters less than 4 μm (physical diameter, about 1.3 μm) will penetrate into the deeper, gas-exchange regions of the lung (the alveoli) (Lippmann 1990a, 1993). However, nasal breathing would result in less penetration of fibers into the lung. The Navy's discussion of the influence of fiber length on regional fiber deposition should cite the work of Sussman et al. (1991a, 1991b), who showed that fiber length is an important determinant of fiber deposition in lung conductive airways, particularly for fibers longer than 10 μm. Also, the Navy's brief review of the biological effects of fibers longer than 15 μm should cite the stronger association of lung-cancer incidence with the number concentration of fibers longer than 10 or 20 μm with lung cancer incidence, rather than of fibers longer than 5 μm (Lippmann 1994). The subcommittee suggests that the association of lung cancer with fibers longer than 10 or 20 μm might stem from interception and preferential deposition of fibers longer than 10 μm at bifurcations of large airways. Fiber length might also be associated with disease because longer fibers, in general, are more biopersistent, more toxic to cells, and more mutagenic to chromosomes.

The Navy's review of relevant literature on in vivo fiber dissolution is fairly thorough for work conducted before the middle 1990s but is insufficiently interpretive. The review should emphasize more-recent work, particularly analyses that have used the available literature to construct and validate predictive models of biological response based on fiber length, diameter, and biopersistence. Some recent extremely relevant papers not cited by the Navy are those of: Collier et al. (1997), Bernstein et al. (1996), Eastes and Hadley (1995, 1996), Kamstrup et al. (1998), Hesterberg et al. (1998b, 1999); McConnell et al. (1999); Zoitos et al. (1997); and Searl et al. (1999). Those papers highlight the critical importance of fiber dissolution and lung biopersistence in determining the extent of the risks of lung fibrosis and cancer. They strongly suggest that risks posed by exposure to MVF are more closely related to lung biopersistence than to the number concentration of long fibers or to the gravimetric concentration of inhalable dust. Therefore, a limit of 1 or 2 fibers per cubic centimeter might be highly appropriate for MVF with a given range of biopersistence, but too conservative for less-biopersistent fibers and insufficiently protective for more-biopersistent fibers.

ASSESSING BIOPERSISTENCE

The primary determinants of biopersistence in the lung are the sites of fiber deposition and the extent of clearance. The dimensions of a fiber determine its initial deposition, which, with its durability in the lung, controls its biopersistence and potential translocation. MVF that deposit in intact conducting airways do not appear to cause a significant pathogenic response, because the mucous coating and rapid clearance via the mucociliary escalator prevents interaction between the MVF and the epithelial lining. Airways that are denuded of an epithelial covering, however, because of confounders such as smoking—could conceivably be vulnerable; fibers might then persist in a location not normally susceptible to fiber-related pathogenicity.

The initial response to MVF in noncompromised lung (such as that of a nonsmoker) is restricted to the alveolar region and is driven by the physicochemical nature of the fiber. Surface properties of fibers can be important for assessing their pathogenicity (Fubini et al. 1998; Kane 1996), but this appears to be of less importance for vitreous (amorphous) fibers than for natural fibers that have a crystalline nature, such as asbestos (Heaney and Banfield 1993).

The typical lung response to a fiber is characterized initially by the recruitment of neutrophils and resident macrophages, which attempt to phagocytize or destroy the offending material, as the lung would any foreign body. The initial response (within hours) in the lung is inflammatory and, for the most part, is restricted to the gas-exchange area of the lung, the site of initial deposition (for example, the bronchoalveolar junction and proximal alveolar ducts and alveoli) (Brody et al. 1985; Warheit 1994). If the fiber remains in the lung long enough, and if the dose is sufficient, cascade of events can ensue, including clustering of macrophages, interstitial and pleural fibrosis, and in some instances neoplasia.

With regard to fiber dimensions, it has been shown in both rodents and humans that most short fibers (less than 5 μm in length) (Bernstein et al. 1996) and nonfibrous particles (Lehnert et al. 1990; Raabe et al. 1986) deposited in the conductive airways of the lung are removed via the mucociliary apparatus within 24-72 hr. The half-life of particles and short fibers deposited in the gas-exchange region is 60-90 days in rats, and up to 10 times greater in humans, depending on their clinical condition. Several animal studies have demonstrated that the inflammatory response stimulated by particles and short fibers resolves once exposure ceases and the offending material is removed from the lung (Hesterberg et al. 1993; Mast et al. 1995a; McConnell et al. 1994, 1999). Long fibers deposited in the gas-exchange airways require macrophage clearance, whose efficiency is dictated by fiber length. The biological explanation of that observation is that fibers shorter than the diameter of a macrophage (rat,10-13 μm; human,14-21 μm) (Crapo et al. 1983; Krombach et al. 1997; Sebring and Lehnert 1992; Stone et al. 1992) can be phagocytized and removed from the lung. Fibers that are too long to be entirely engulfed by a macrophage cannot be removed and therefore have a greater chance of interacting with the epithelium and tissues under it with resultant pathology.

A fiber must reside in the lung for a sufficient time to elicit a chronic response, such as fibrosis or neoplasia. The residence time of a fiber is dictated by its durability. The durability of an MVF depends on its in vivo solubility, a function of its chemical characteristics, length, and diameter and of the microenvironment of the lung. MVF biodegrade in the lung primarily via dissolution. However, a fiber whose length would normally keep it from being removed efficiently, does not have to be completely dissolved to be removed. For example, if a fiber is weakened at any point in its length, it can break into fragments, which can then be removed via macrophages. MVF are amorphous and always break transversely in contrast with some crystalline fibers, such as asbestos, which tend to split longitudinally (Bellman et al. 1987; Hesterberg et al. 1998a, 1998b). Therefore, as MVF dissolve or break, their removal from the lung is likely to be expedited and their adverse effects mitigated compared with fibers such as asbestos.

In vitro techniques to determine fiber durability at different pH values and in vivo tests to determine overall biopersistence in the lung directly have been developed. Short fibers that are ingested by alveolar macrophages encounter an acidic pH of 4.5-5 in the phagolysosomes (Lundborg et al. 1995), whereas longer fibers, not phagocytizable by alveolar macrophages, are subjected to the extracellular-fluid pH of about 7.4. Therefore, in vitro tests to measure the leaching of specific fiber constituents into the dissolution medium are generally performed at pH values of both 4.5 and 7.4 to simulate the intracellular and extracellular milieus, respectively (Potter and Mattson 1991). Although acellular in vitro techniques measure only dissolution rates of the fibers, in vivo methods measure the overall retention of fibers in the lung, including dissolution rates and mechanical clearance. A comparison of results from in vitro and in vivo techniques is summarized in Table 4-1 for amosite and crocidolite asbestos and several MVFs. Maxim et al. (1999a) presented similar data showing the in vitro dissolution rates of different vitreous fibers plotted against the retention half-times in the lung after inhalation. They demonstrated that there is excellent correlation ($R^2 \approx 0.90$) between in vitro and in vivo measurements (see Figure 4-1). However, in certain cases, current methods of measuring in vitro dissolution rates might not be good predictors of in vivo behavior of fibers longer than 20 μm, that is, fibers that are not phagocytized by alveolar macrophages. More research is needed to optimize and standardize in vitro dissolution methods.

MEASURING BIOPERSISTENCE

The biopersistence of a given MVF is the result of several physical and biological factors, a combination of which are required for pathogenicity (Barrett et al. 1989). In vivo and in vitro studies have been used to determine biopersistence of MVF; in vivo studies are preferred but are expensive and time-consuming.

TABLE 4-1 In Vitro Solubility and In Vivo Biopersistence of Asbestos and Various Types of MVF in Lung

Fiber class	Fiber type and use[a]	K_{dis}[b] at pH 7.4	Time to dissolve[c], days	WT ½,[d] days	Ref.
Asbestos	Amosite	<1	>5,000	~400	1
Asbestos	Crocidolite	<1	>5,000	~800	1
RCF	High temperature applications	3	~1,700	~50	1
Other-RCF substitute	Magnesium silicate	>150	~35	~6	2
Glass	MMVF 32, specialty applications	9	~550	~80	1
Glass	MMVF 33, filtration	12	~400	~50	1
Glass	MMVF 11, building insulation	100	~50	~10	1
Glass	A, building insulation	250	~20	~4	1
Glass	MMVF 10, building insulation	300	~20	~35	1
Glass	P, building insulation	>500	<10	~5	1
Glass	C, building insulation	>500	<10	~4	1
Glass	B, building insulation	>500	<10	~2	1
Rock	MMVF 21, building insulation	20	~250	~60	1
Rock	MMVF 34[e], building insulation	60	~80	~5	1

Rock	F, building insulation	160	~30	~9	1
Rock	G, building insulation	210	~20	~5	1
Rock	H, building insulation	270	~20	~13	1
Experimental	J, building insulation	170	~30	~10	1
Rock	O, building insulation	>500	<10	~6	1
Slag	MMVF 22, ceiling tile	400	~15	~9	1

References: (1) Hesterberg et al. 1998a, and (2) Zoitos and Boymel 1999.

[a] MMVF 10, 11, etc., and A, B, etc., are designations that have appeared in the literature.
[b] Dissolution rate of fibers in vitro, k = ng/cm^2-hour.
[c] Time it would take a 1.0-μm-diameter fiber to dissolve completely.
[d] Weighted clearance half-time.
[e] Example of new class of rock wool fibers. They have high dissolution rates at low pH and rapid removal from the lung after inhalation. Estimate of dissolution rate at pH 7.4 is very uncertain.

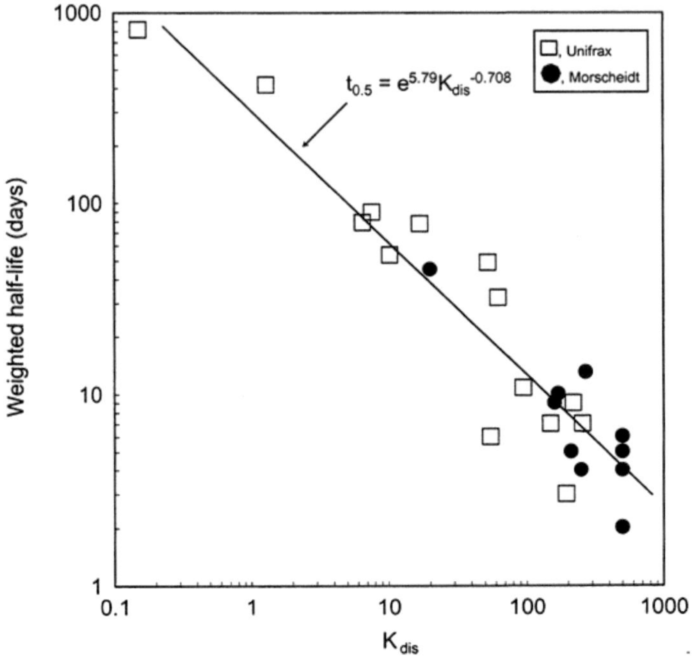

FIGURE 4-1 Relation between in vivo weighted half-life of fibers in short-term inhalation experiments and K_{dis}. Source: Adapted from Maxim et al. (1999a).

The impact of the aggregate effects of dissolution, breakage, and removal from the lung has been evaluated in vivo. Initial studies to investigate these phenomena were "add-ons" to standard chronic rodent bioassays. In long-term studies, animals (usually rats) were exposed to concentrations of fibers for 2 years; groups of rats were removed from exposure at various times such as 3, 6, 12, or 18 months—and held unexposed for the remainder of the 2 years. The numbers of fibers remaining in the lungs of the rats exposed for various times were compared (Hesterberg et al. 1993; Mast et al. 1995b; McConnell et al. 1994).

Chronic studies are very costly and time-consuming to conduct, and therefore a simpler method was sought. This has been accomplished through what have been termed "biopersistence" studies (Bellmann et al. 1987). This type of study consists of exposing animals (preferably rats) to a known concentration (number, diameter, and length) of a given fiber type for 6 hr/day for 5 days followed by periodic sacrifices to determine lung fiber burden (Bernstein et al. 1996). Alternatively, rats can be dosed by tracheal instillation with a known amount of fiber in saline, and similar measurements can be made. Although the latter method has been criticized for technical reasons, such as uneven distribution in the lung, it can provide useful comparative information. Findings from both methods are typically presented as weighted clearance half-time (WT½) or as time required to clear 90% of the fibers from the lung (T90) for both WHO fibers[1] and fibers longer than 20 μm. Results of these short-term studies have been compared with the results of long-term toxicity studies in animals to establish a WT½ or T90 and predict what would happen in humans under comparable exposure conditions. The short-term studies have been adopted in Europe as an alternative to long-term carcinogenicity bioassays (EC 1999).

The results of the studies show that various types of MVF differ in their biopersistence (Table 4-1). For instance, Hesterberg et al. (1998a) have reported that the WT½ of MMVF 11 (a standard type of fiberglass insulation) is about 10 days, that of MMVF 33 (a specialty type of fiberglass used for filtration) is about 50 days, and that of amosite asbestos is about 400 days. The T90s of the same fibers were 38, 240, and 2,095 days, respectively. Those findings underscore the importance of biopersistence in explaining the differential pathogenicity of the same MVF, as was seen by McConnell et al. (1999) in a standard lifetime inhalation study in hamsters. McConnell et al. showed that at equivalent exposures, MMVF 10a (similar to MMVF 11) caused only an inflammatory change, MMVF 33 caused a moderate amount of pulmonary and pleural fibrosis and a single mesothelioma, and amosite asbestos caused severe lung and pleural fibrosis and a high incidence of mesotheliomas.

[1] WHO fibers are defined as having a length greater than 5.0 μm, a diameter less than 3.0 μm, and an aspect ratio equal to or greater than 3:1 (WHO 1985).

The postexposure retention kinetics of deposited fibers can generally be described by a biexponential curve, with a fast phase reflecting fiber clearance from the conducting airways and potential clearance of fiber fragments and a slow phase reflecting—in the case of long fibers (longer than 20 μm)—the elimination of nonphagocytized fibers due to dissolution and breakage. European regulations have adopted the use of a WT½ to describe this clearance behavior (EC 1997) but a numerically calculated half-time may also be used (Miller 1999). However, depending on the fraction of fibers cleared in the fast phase, the fast value could reduce the WT½ to such a degree that the difference between a soluble (low-biopersistence) fiber and a poorly soluble fiber becomes minimal; that is, the power of differentiating between the two might be low. Thus, although the use of WT½ will discriminate different types of fibers, the results could be misleading because both fiber clearance and biopersistence must be considered in disease causation. Therefore, the use of T90 or WT½ values in risk management needs to be evaluated carefully.

In vitro dissolution studies have been proposed as surrogates to predict in vivo solubility of MVF and other fibers (Eastes and Hadley 1996; Zoitos et al. 1997). The rationale is that they are inexpensive and fairly rapid and have been shown to correlate fairly well with the in vivo biopersistence studies discussed above (Eastes et al. 2000). The technique involves placing fibers in something as simple as physiological saline or a solution that mimics what is found in the lung, such as Gamble's solution, and then measuring dissolution of the fiber over time. Dissolution can be measured by determining the dissolution-fluid composition or the weight of the remaining fibers (Potter and Mattson 1991). Such studeis are typically conducted at a pH of 7.4 (the pH of lung fluids), but they can also be performed at a pH of approximately 5 (the pH in a macrophage). The dissolution rate (K_{dis}) is expressed as nanograms per square centimeter per hour. The results of these studies, like those of in vivo studies, show that MVF vary markedly in solubility within a fiber type and between types (Table 4-1).

Another innovative method that has been used to investigate biopersistence is confocal laser scanning microscopy (CLSM), which allows fibers to be directly visualized in their exact location in the lung and measured in situ without disturbing the lung by fixation or cutting. Rogers et al. (1999) used CLSM to examine portions of lungs from hamsters that had been exposed for 13 weeks, 5 days/wk, 6 hr/day to MVF in a study conducted by McConnell et al. (1999). Rogers et al.

showed that at the end of the exposure period the lengths of MMVF 10a, MMVF 33, and amosite fibers in the lung were reduced by 64.2%, 44.2%, and 8.7%, respectively, in comparison with the original aerosolized-fiber lengths. McConnell et al. (1999), showed that after a single 6-hr exposure in hamsters, almost twice as many WHO amosite fibers and approximately equal numbers of fibers longer than 20 μm were deposited in the lung. The results of CLSM constitute additional evidence that differential shortening occurs in fibers retained in the lung. However, the value of CLSM is limited as it can only be used to visualize to a depth of a few millimeters in the lung, whereas more-centralized locations cannot be visualized without cutting the lung, which would diminish the advantages of the technique. At present, CLSM should be viewed as a research tool rather than as a means of assessing biopersistence definitively.

CONCLUSIONS

The potential hazards posed by a given MVF is directly related to its ability to persist in the lung long enough to cause chronic disease. This persistence has been termed biopersistence. Research has shown that persistence in the lung is directly related to the chemical composition and dimensions of fibers. Biopersistence can be measured with long-term and short-term studies. It has also been proposed that simple in vitro fiber solubility studies can reasonably predict what will happen in vivo. Most of the fibers that are of concern to the Navy have been investigated with one or more of these techniques, and review of the resulting data can be valuable for estimating the hazards potentially associated with the fibers.

5

TOXICOLOGICAL STUDIES

N support of its selection of an occupational exposure standard of 2 fibers/ cm^3 for manufactured vitreous fibers (MVF), the Navy reviewed much of the available toxicological literature published before 1997. It presents this information as taken from in vitro studies, epidemiological studies (discussed in the next chapter), and animal toxicity studies. The animal toxicity studies are grouped by fiber type and route of administration. The published literature cited by the Navy are current through 1997. However, some relevant toxicological material published since the Navy's 1997 report might inform the Navy's selection of an occupational exposure limit.

The Navy correctly notes that inhalation studies have yielded the most relevant data, as they were conducted using the route of administration that most closely mimics expected human exposures. It also acknowledges the controversy with regard to some aspects of animal toxicity testing of MVF, including the validity of intrapleural and intraperitoneal administration. The subcommittee agrees with the Navy that the route of administration is one of the most controversial aspects of toxicity studies of MVF. Although the Navy does mention some of the controversy and limitations of the toxicity studies it reviewed, it does not elaborate on the limitations of the noninhalation studies. In general, an assessment of the toxic effects of inhaled fibers requires consideration of both the animal model and the fibers' characteristics, including its dimensions, durability, biopersistence, and surface characteristics.

Inhalation, intratracheal instillation, and intracavitary injection studies in animals have been used for estimating the biopersistence and hence potential toxicity and carcinogenicity of inhaled MVF in humans. Each kind of study has advantages and limitations, as discussed in McClellan et al. (1992) and McConnell (1995) and briefly presented below.

INHALATION STUDIES

Experimental data are essential for providing basic information on the physiological and pathophysiological pulmonary responses to inhaled particles. Because various rodent species respond differently to selected inhaled materials, it is essential to consider numerous factors—such as anatomy and deposition patterns, physiology and macrophage clearance efficiency, biochemistry and inflammation and fibrogenic potential—when extrapolating the results of animal inhalation studies to humans. Therefore, knowledge of morphological and functional pulmonary characteristics is essential for full understanding of structure-function relationships among species but it is also necessary if one is to develop accurate risk estimates with regard to the toxicity of inhaled particles in exposed humans.

Several rodent species are commonly used in particle and fiber inhalation-toxicity studies designed to simulate human exposures and to evaluate lung responses to inhaled dusts. But experimental animals and humans differ with respect to lung anatomy and physiology and these differences influence particle deposition and corresponding lung-clearance responses. For example, humans have relatively symmetrical dichotomous airway branching that favors concentrated deposition on branch points, or bifurcations; rodents have highly asymmetric, monopodal branching that theoretically should reduce the tendency for concentrated deposition. Distal airways are fundamentally different between humans and rodents: humans have several generations of nonrespiratory bronchioles and three generations of respiratory bronchioles and alveolar ducts; guinea pigs and hamsters have poorly develop respiratory bronchioles, and mice and rats generally lack them. Humans and rodents have different pleural tissue anatomy. And rodents are obligate nasal breathers, whereas humans can favor oral breathing while speaking or during strenuous activity, thus permitting enhanced particle penetration to the lungs.

Several studies have used rats and hamsters as the primary species for assessing the chronic effects of inhaled fibers (Mast et al. 1994; Mast et al. 1995a; McConnell et al. 1999). Some have demonstrated clear interspecies differences in lung-tumor and mesothelioma responses to inhaled synthetic fibers. Rats appear to be more likely to develop lung tumors after exposure to refractory ceramic fibers (RCF) than hamsters, which have greater sensitivity for developing mesotheliomas (Mast et al. 1994; McConnell et al. 1994). Hamsters appear to be resistant to the development of lung tumors after chronic exposure but appear to be extremely sensitive to mesothelioma induction after exposure to selected fiber types. Because few chronic fiber inhalation studies of appropriate reference materials have been conducted in hamsters, it is difficult to determine whether the hamster is a relevant model for humans. Similarly, interpretations of lung-tumor response in chronically exposed mice are difficult because of the high incidence of spontaneous lung tumors.

Nevertheless, mammalian inhalation tests have some obvious advantages over other tests. The route of exposure is similar to that in humans, and the exposure to fibrous materials is directed to the intact pulmonary system, including all natural defense mechanisms. In rats, the incidences of fibrosis, lung cancer, and mesothelioma after exposure to asbestos are comparable with those in humans (Warheit and Hartsky 1994).

Disadvantages of animal inhalation studies include species differences in respiratory anatomy and function noted above, and species-specific pathological responses in control and treated animals (especially, in the latter, to exposures that result in overloading of the animals' capacity to clear deposited particles and fibers). Animal inhalation studies for fiber toxicity screening tend to be time-consuming, are expensive, and cannot necessarily elucidate the details of cellular and molecular events.

Despite the limitations, a panel of the World Health Organization (WHO) has concluded that inhalation studies constitute the best available laboratory model for assessing the human health risks posed by exposures to fibers (McClellan et al. 1992; WHO 1992).

Subchronic and chronic inhalation tests are typically used to study health effects and dose-response relationships. Recently, short-term inhalation studies (about 1-day to 2-weeks) with extended followup have been used to study biopersistence, cellular reactions, proliferative reactions, and repair and clearance mechanisms. For studying biopersistence of MVF in this fashion, methods for digesting the lung must be validated. Some techniques for validating the methods, such as low-temperature ashing and digestion with strong acids or bases, have limitations. For instance, low-temperature ashing can make the fibers brittle or artifi cially break them, and digestion with strong acids or bases can destroy retained fibers or alter their composition. Therefore, alternative digestion techniques that preserve each fiber type need to be developed.

That physical characteristics of the fibers, such as fiber dimensions, play an important role in the pathogenesis of fiber-associated lung disease was demonstrated clearly by Davis et al. (1986), who compared the effects of short and long amosite asbestos fibers at equivalent mass concentrations. Rats were exposed for 1 year by inhalation to aerosols of specially prepared short amosite asbestos fibers (shorter than 5 μm) or long amosite asbestos fibers (longer than 20 μm); the two preparations were derived from the same source and at equivalent gravimetric concentrations. As a result, rats were exposed to greater numbers of short fibers than long fibers. After exposure, no significant histopathological effects were observed in the lungs of rats exposed to the short fibers, but one-third of the rats exposed to the long fibers developed lung tumors. Nearly all the rats exposed to the long fibers also developed diffuse pulmonary fibrosis.

Inhalation toxicity studies in rodents must be extrapolated to humans cautiously. Rats or other rodent species generally are experimentally exposed to high concentrations of preparations of long fibers by enriching the aerosol with the fibers. But, such exposures might not adequately simulate occupational or environmental exposures to lower fiber concentrations or to mixtures of fibers of varied lengths; rather, they are designed to represent a potential worst-case scenario.

INTRATRACHEAL INSTILLATION

Studies that use intratracheal instillation as a route of rodent exposure to fibers are generally regarded as easier and less expensive than inhalation studies. Bolus administration often leads to uneven distribution of fiber-shaped particles throughout the lung and localized overloading (ECETOC 1996). Nevertheless, these types of studies might have value for the initial screening of fibrous compounds. A European Commission (EC) directive for classification and labeling of synthetic mineral fibers (Commission Directive 97/96/EC of December 5, 1997) allows for the use of either the short-term inhalation biopersistence assay or the intratracheal-instillation biopersistence assay in exonerating fibers from classification as a carcinogen. The protocols for performing those tests have been defined by the European Chemical Bureau (EC 1999). The biopersistence protocols are accepted by the EC for interim use and are being validated in a multicenter ring test. The subcommittee believes that instillation tests are useful for ranking the biopersistence of MVF fibers, but their validation will require data on more fiber types than are presently available. Validation should include the deposition of instilled fibrous material into the alveolar regions of the lung, and correlation of biopersistence of the instilled material (as defined by the investigators) with the development of pathological pulmonary effects.

In spite of the limited data available from these studies, intratracheal instillation of materials remains a popular alternative to inhalation exposure for several practical reasons: small quantities of the test compound can be used, thus reducing waste and increasing safety when hazardous materials are being tested; the technique is inexpensive because it does not require expensive exposure chambers and elaborate vapor or aerosol generation apparatus; complex technical support is not necessary for producing and monitoring vapor or aerosol exposures; and high concentrations of particles or fibers can be administered to the respiratory tract at numerous doses with precise control and measurement.

There are also disadvantages to instillation that stem from the differential distribution in the lung of instilled particles compared with inhaled particles. Instilled particles move to the gravity-dependent portions of the lung because the injected material settles, whereas inhaled airborne particles tend to be well distributed throughout the respiratory system, particularly in the small airways. The high local concentration of instillates or their carrier liquids can cause local tissue damage, particularly at high particle or fiber doses. That can lead to local hemorrhage and even death by mechanisms not directly relevant to the study. The acute inflammatory response that develops in response to the high particle burden and liquid suspension of the carrier could actually contribute to the formation of lesions observed in instillation studies. In contrast, the inhalation technique avoids these local and regional overload effects because the lungs of the exposed animals do not receive the full bolus of particles in one dose. Inhalation models best simulate human exposure because only respirable particles reach the lung parenchyma. Instillation techniques, in contrast, can result in the delivery of nonrespirable (large) particles to the alveolar regions, where they normally would not deposit.

Instillation is an acceptable form of dosing in many cases and might be the only practical mean of dosing but it cannot substitute for a properly performed inhalation study. This type of toxicity study has been used in fiber clearance and biopersistence studies.

INTRACAVITARY INJECTION

Intracavitary tests, such as intraperitoneal and intrapleural fiber-injection studies, are conducted primarily in rats. In many cases, rats are given abdominal or pleural injections of a bolus that contains from 10^6-10^9 fibers and are then evaluated at the end of their lifespan or when a tumor is identified. These tests are known to produce a high incidence of mesotheliomas. Intracavitary models have been advocated as relatively inexpensive and highly sensitive tests to predict the carcinogenicity of fibers (Stanton et al. 1981; Pott 1980). However, the route of administration bypasses all natural defenses, and a single dose (or a few repeated doses) early in life might not necessarily produce the physiological responses that would be observed at lower doses and longer exposures. There is considerable concern that intracavitary models can give false-positive results, even for the prediction of mesothelioma risk, and there is no agreement over their predictive value for lung cancer. The subcommittee agrees with a WHO scientific panel's conclusion that the intraperitoneal model should not be used for quantitative risk assessment or for comparing relative hazards posed by different fibers (WHO 1992).

CONCLUSIONS

It appears reasonable to conclude that extrapolations from animal toxicity data to humans for MVF can best be made when experimental animals are exposed to fibers via inhalation. Studies using instilled doses are valuable insofar as they provide a rough estimate of the pulmonary toxicity of materials, but they should not be used for hazard assessments when setting exposure limits.

Intracavitary exposures, via either intraperitoneal or intrapleural injections, can produce a high incidence of mesotheliomas. Such exposures have been advocated as relatively inexpensive and highly sensitive tests to predict the carcinogenicity of inhaled fibers (Pott et al. 1989). However, this route of administration bypasses all natural pulmonary defenses, and the single dose (or a few repeated doses) is not physiologically based and can create an overload in the peritoneal or pleural cavity. Intracavitary tests can also yield false-positive results, for the assessment of lung cancer and mesothelioma risks. The WHO consultation (WHO 1992) concluded that the intracavitary model should not be used for quantitative risk assessment or for hazard evaluation of fibers.

Of the three types of tests that can be used to screen for fiber toxicity—inhalation, instillation, and intracavitary—one might be more advantageous than another. Intracavitary tests are not recommended because of the numerous deficiencies discussed above. Results of instillation studies are qualitatively similar to those of inhalation studies (Henderson et al. 1995) and are adequate for short-term estimates of toxicity and fiber-clearance studies, but they cannot substitute for inhalation models for setting dose levels. Short-term inhalation testing should be used for estimating toxicity, evaluating mechanisms, and setting doses for subchronic or chronic inhalation studies. With regard to the latter goal, it is likely that the data generated from short-term inhalation tests could be used to set dose levels for 90-day inhalation studies, thus obviating costly 2-week or 28-day dose-setting inhalation studies.

6

EPIDEMIOLOGICAL STUDIES

HE Navy Environmental Health Center prepared a comprehensive review of the epidemiological literature in support of its proposed occupational standard for manufactured vitreous fibers (MVF) (NEHC 1997b). There have been well over two dozen publications of epidemiological studies of MVF production facilities in North America and Europe that have examined large cohorts of workers. Those studies have used both historical cohort mortality and case-control study designs. Interest has centered primarily on lung cancer, although other cancer sites and causes of death have been considered. Lee et al. (1995) have reviewed the published work with respect to respiratory system cancers associated with MVF.

The Navy's documentation in *Man-Made Vitreous Fibers* evaluated data primarily from the three large cohort studies conducted in Europe, the United States, and Canada. Since publication of the Navy's report, several new epidemiological studies (Chiazze et al. 1997; Watkins et al. 1997; Hansen et al. 1999) and updated analyses (Boffetta et al. 1997; Boffetta et al. 1999; Chiazze et al. 1999; Marsh et al. 1996) of earlier studies of MVF have been published. The subcommittee suggests that the Navy update its report to discuss the strengths and limitations of the recent cohort and case-control studies so that their results can be adequately assessed for the Navy's needs. More recent studies that examined the health effects of refractory-ceramic-fiber (RCF) production workers might be able to account for potential biases and confounders such as smoking that limit the older epidemiological studies of MVF. In analyzing and interpreting results from these studies, it is imperative that the Navy assess their underlying limitations.

MANUFACTURED VITREOUS FIBERS

Chiazze et al. (1997) conducted a historical cohort mortality study of a continuous-filament glass-fiber manufacturing plant in Anderson, South Carolina. The investigation was undertaken in light of evidence of a statistically significant increase in the proportionate mortality ratio for lung cancer among white men at the plant. A nested case-control study of lung-cancer deaths among white men was incorporated into the study design. Information on sociodemographic factors—including smoking, alcohol consumption, and medical history—was obtained through an interview survey; exposure assessments for respirable glass fibers, total particles, asbestos, RCF, respirable silica, formaldehyde, total chrome, and arsenic were gathered through reconstruction of historical data. Respirable glass fibers were produced at the plant only in 1963-1968. Lung-cancer odds ratios (ORs) among workers exposed to respirable glass fibers were below unity, as were ORs for exposures to asbestos, RCF, respirable silica (except for the lowest exposure level), total chrome, and arsenic. Those results suggest that none of the plant exposures resulted in an increase in lung-cancer risk for this population, although the lung cancer standardized mortality ratio (SMR) is slightly increased (observed, 47; SMR, 126). The authors conclude that neither glass fibers nor any of the substances investigated as part of the plant environment were associated with an increase in lung-cancer risk in the study population.

In a separate cohort study conducted at the same continuous-filament glass-fiber plant in Anderson, South Carolina, Watkins et al. (1997) examined mortality in women and minority groups. The study population consisted of 1074 white women, 130 black women, and 494 black men who worked at least one year from the opening of the plant in 1951 through 1991. The women made up the largest cohort of white women assembled to date in a wool or continuous-filament glass-fiber manufacturing facility, and this was the first study of black men and women in the MVF industry. Over 95% of the women and minority-group members included in the study held production positions in the plant. Results of the analyses demonstrated that there were no significant excesses or deficits in mortality among white women by cause of death—including cancer—compared with national mortality, except for motor vehicle accidents. Among black men, SMRs for all cancers combined were below unity on the basis of local and national standards; and lung-cancer SMRs were below unity among white women and black men. However, the authors mentioned that the statistical power of the study was low and possibly not adequate to detect risks of the magnitude typically of interest in studies of men.

Hansen et al. (1999) conducted a cross-sectional survey of 235 Danish rock-wool production workers with at least 5 years of exposure in 1955 and 1987. The reference group consisted of 243 nonexposed subjects randomly sampled from the general population in the same area. Respiratory health was assessed by questionnaires, spirometry, and measurement of diffusion capacity of the lungs. The authors determined, after adjusting for age and smoking, that there was no association between MVF and respiratory symptoms as reported in the questionnaires. Self-reports of emphysema were more common in exposed workers (3.8%) than in nonexposed workers (0.9%) (RR, 4.5), but only nine exposed workers and two nonexposed workers reported emphysema. Of the lung-function parameters measured, only the lower FEV_1/FVC ratio (ratio of forced expiratory volume in 1 second to forced vital capacity) was significantly associated with MVF exposure after adjusting for age, height, and smoking. The prevalence of airflow obstruction was greatest among exposed workers with more than 40 pack-years of smoking; this suggested a synergistic effect between smoking and MVF. A similar interaction between smoking and MVF was reported by Trethowan et al. (1995) in ceramic-fiber workers.

Several updated analyses of data sets have been published recently. Marsh et al. (1996) published an update of the 1946-1989 mortality experience of the rock- and slag-wool subcohort of the North American Insulation Manufacturers Association (NAIMA) cohort. (The NAIMA cohort comprises production and maintenance workers in 17 of the oldest and largest MVF plants in the United States.) The study authors reported a statistically significant increase in lung cancer in the subcohort (observed, 70; SMR, 129). However, as in all the historical cohort mortality studies of MVF, SMRs were not adjusted for the possible effect of confounding by cigarette smoking. There were no statistically significant increases in the SMRs for any of the remaining cause-of-death categories except nephritis and nephrosis (observed, 12; SMR, 204). Exposures were estimated quantitatively for respirable fibers, asbestos, formaldehyde, and silica and qualitatively for arsenic, asbestos, asphalt, polycyclic aromatic hydrocarbons, phenolics, radiation, and urea. The exposure estimates were used in relative-risk regression modeling, which yielded no consistent evidence of an association with any of the respirable-fiber measures considered. The findings were corroborated in a nested case-control study that adjusted for smoking.

Boffeta et al. (1997) published results of an updated mortality followup of a cohort of MVF production workers in Denmark, Finland, Norway, Sweden, the United Kingdom, Germany, and Italy (Simonato et al. 1987). The population studied consisted of people employed in 13 factories in 1933-1950 (when production was started), and followed through 1990, 1991, or 1992, depending on the country. Historical exposure information on MVF was based on production periods when workers were employed in plants (Simonato et al. 1987); no quantitative estimate of MVF exposure was available. Among specific cancer causes of death, SMRs were statistically significantly increased only for lung cancer in the rock- and slag-wool subcohort (observed, 97; SMR, 1.34), with 42 of the deaths in workers in one factory in Germany, and the glass-wool subcohort (observed, 140; SMR, 1.27), with 109 of the deaths in workers in one plant in the United Kingdom. The SMR was increased for workers with less than 1 year of MVF employment (SMR, 1.48) and with more than 1 year (SMR, 1.29). There were no statistically significant increases for cancer in the continuous-filament subcohort (Boffeta et al. 1997). The authors conclude that the results were not sufficient to find that the increased lung-cancer risk is specifically related to exposure to MVF. However, they note that insofar as respirable fibers were an important component of ambient concentrations in the workplace, these fibers might have contributed to the increased risk.

In a further followup analysis of the rock- and slag-wool and glass-wool cohorts in Denmark, Norway, Sweden, and Finland, Boffeta et al. (1999) found an increased standardized incidence ratio of 1.15 for lung cancer in the combined cohorts. However, it could not be correlated with fiber exposure. Time since first employment and duration of employment were used as surrogate measures when no other exposure measures were available. It is important to note that none of the SMRs in the European study was adjusted for the possible confounding effect of cigarette smoking; exposures to other carcinogens, polycyclic aromatic hydrocarbons, and asbestos might also have occurred.

In another European study, Sali et al. (1999) investigated the nonneoplastic mortality of rock- and slag-wool, glass-wool, and continuous-filament workers in MVF factories in Denmark, Finland, Norway, Sweden, the United Kingdom, Germany, and Italy. Exposure to MVF was based on various measures, including time since first employ ment, duration of employment, and technological phase at first employment. The authors determined that mortality from nonneoplastic diseases did not appear to be related to employment in the MVF industry. However, they believe that an increased risk of death from nonmalignant renal diseases among rock- and slag-wool workers with increasing duration of employment and an increased risk of ischemic heart disease in rock- and slag-wool and continuous-filament workers with more than 30 years since first employment warrant further investigation.

Recently, Chiazze et al. (1999) conducted a case-control study based on the Owens Corning Surveillance System to investigate the question of whether there is an association between silica or respirable glass fiber exposure and mortality from nephritis and nephrosis among workers in fibrous glass wool manufacturing facilities. Quantitative estimates of exposure to silica and respirable fibers were obtained through historical environment reconstruction. Information on sociodemographic factors—including education, marital status, income, drinking, and smoking—was gathered through interview surveys. The study authors found no consistent relationship for silica or respirable fibers when the analysis was based only on underlying cause of death. When the analysis was based on both underlying and contributing causes of death, the odds ratios for respirable fibers were below unity for both the lowest exposure (OR, 0.81; 95% CI, 0.30-2.13) and the highest exposures (OR, 0.53; 95% CI, 0.22-1.31). The results were similar for silica: ORs were below unity except for the highest exposure, for which the OR was 1.04 (95% CI, 0.24-4.46). The authors state that although the results do not prove the lack of association between nephritis and nephrosis and glass-fiber or silica exposure, they do not support such an association. However, the findings do suggest that all information on the death certificate, not just the underlying cause, should be used to capture the most accurate picture of renal disease.

Those cohort and case-control studies have contributed to expanding the existing database on effects of exposures to MVF. However, the confounding factors and limitations in the exposure information that affect the interpretation of the results must be considered. Typically, exposures in the plants have been relatively low—average plant concentrations were reported to be 0.005-0.292 f/cm^3 for fibrous glass and 0.194-0.426 f/cm^3 for mineral wool (Marsh et al. 1990)—and workers were exposed to multiple types of fibers, and to other potential carcinogens at the plants, including asbestos. Because of the retrospective nature of the studies, direct exposure measurements typically do not exist, and exposure information must be reconstructed from employment histories and plant records.

Inability to control for potential confounders, in particular cigarette smoking, is especially important because smoking is such a strong predictor of respiratory system cancer. For instance, Chiazze and Watkins (1994), in their reanalysis of data from the Owens Corning plant in Newark, Ohio, demonstrated how important it is for lung-cancer SMRs based on national data to take the confounding effect of smoking into account. They derived an overall estimate of the average prevalence of ever having smoked cigarettes in 1940-1980 (the followup period of the Newark cohort) for white men in the United States at least 20 years old from the National Health Interview Survey special smoking supplements for 1978-1980. The estimate was based on survey participants who were self responders with known age, race, sex, smoking status, and age when they began smoking. After adjustment for the confounding factor of cigarette smoking, the adjusted SMR was found to be below the level of significance and quite similar to the SMR obtained by using local rates.

Several review articles (McClellan et al. 1992; Lee et al. 1995) have recommended that future studies provide information on cigarette smoking and other workplace carcinogens because future studies would not provide useful information without such data. The issue of statistical significance is moot unless potential extraneous factors are taken into account. A statistical-significance test (or any mathematical model) cannot compensate for lack of appropriate data on exposures and confounders (Chiazze and Watkins 1994). As Lee et al. (1995) state in their review article, "available data indicate that among those occupationally exposed, glass fibers do not appear to increase the risk of respiratory system cancer. Exposure to rock and slag wool may increase the risk of such cancers; however, the data do not convincingly prove that this association is causal. In addition to inconsistent patterns from cumulative exposure, latency, and duration of employment, the potential for confounding by cigarette smoking and other workplace carcinogens exists in these studies of rock and slag wool fibers."

REFRACTORY CERAMIC FIBERS

In the last several years, there has been increased use of RCF. There is also a growing database of studies examining the health effects on workers employed in the manufacture of RCF. These studies, of which several are cross-sectional or prospective analyses (Lemasters et al. 1998; Lockey et al. 1998), might be better able to account for confounding factors, including cigarette smoking, in their analyses.

Lockey et al. (1996) reported the results of a retrospective cohort study and a nested case-control study designed to evaluate chest radiographs of 652 workers involved in the manufacture of RCF at two sites. A latency-validation review was also conducted in which historical chest xrays were examined to determine whether there was a biologically plausible latency between initial RCF exposure and first appearance of pleural plaques. The study included men and women employed at the time of interview—October 1987-December 1991—and former employees who had been employed for at least 1 year. The cohort study demonstrated an association between exposures to RCF and the occurrence of pleural plaques as measured by three exposure indexes: time from beginning RCF production job, years of RCF-production employment, and cumulative exposure in fiber-months per milliliter. The highest prevalence of pleural plaques was observed in the highest-exposure groups for each exposure index. The nested case-control study, which controlled for the possibility of past asbestos exposure as a potential confounding factor, corroborated the findings of the cohort study. The latency-validation review of historical xrays confirmed that a biologically plausible period had elapsed between initial exposure and development of pleural plaques. The authors concluded that "the cohort and nested case-control studies and the latency validation review support a causal hypothesis for RCF exposure and the development of pleural plaques."

In another study, Lockey et al. (1998) conducted a prospective analysis to evaluate the relationship between RCF exposure and pulmonary-function changes in 361 men employed at five RCF manufacturing plants in 1987-1984. Workers included in the analysis provided at least five pulmonary-function tests. The exposure-response relationship was modeled with two exposure variables: years in a production job and cumulative fiber exposure (fiber-months per milliliter). Analyses by Rice et al. (1997) reconstructed exposure estimates for workers in two plant locations, calculating cumulative RCF exposures from pre-1987 exposure data and current exposure data. However, cumulative exposures could not be calculated for the other three plants, because pre-1987 exposure data were not available. The authors determined that employment in RCF production and cumulative RCF exposure were not associated with a decrement in longitudinal FVC or FEV_1 values from 1987 through 1994. However, a cross-sectional analysis of the initial pulmonary-function test results found that production workers exposed for more than 7 years had lower FVC and FEV_1 than baseline nonproduction workers; the decrease might have been caused by earlier exposure to RCF concentrations that were greater than those in 1987-1994. Those results correspond with the historically greater exposure levels in the 1950s—an estimated maximal exposure of 10 fibers/mL, compared with recent exposures that ranged from below the detection limit to 0.66 fiber/mL (Rice et al. 1997). The authors state that "the decrease in RCF exposure levels over the last 10 years through engineering and work practice changes has reduced any detectable continued effect of RCF exposure on FVC and FEV_1." They plan to study workers with fewer than five spirometry tests to further assess participation bias because it was noted that smokers with reduced lung function were less likely to have produced at least five spirometry tests which were required for inclusion in the longitudinal analysis.

Lemasters et al. (1998) also conducted an industrywide pulmonary-morbidity study to evaluate the respiratory health of employees engaged in the manufacture of RCFs at five U.S. sites in 1987-1989. Of the 753 eligible employees, 742 provided occupational histories and completed the American Thoracic Society respiratory-symptom questionnaire; and 736 of the 742 also underwent an initial pulmonary-function test. Exposure to RCF was assessed by classifying workers as production or nonproduction employees and computing the duration of time the former spent in production. The results of the cross-sectional study demonstrated an increase in respiratory symptoms in production versus nonproduction workers and declines in FVC and FEV_1 for each 10 years of employment in excess of the declines attributed to smoking, age, and other factors. Among current and past male smokers, a significant decline in FVC was associated with employment in the production of RCF. A significant decline in FEV_1 in men who were current smokers was associated with work in RCF production A significant decline in FVC was seen in production women but not production men who had never smoked. The results suggest that there might be important sex differences in responses to occupational or environmental exposures.

CONCLUSIONS

In a review of the published epidemiologic literature with respect to respiratory system cancer, Lee et al. (1995) concluded that "available data indicate that among those occupationally exposed, glass fibers do not appear to increase the risk of respiratory system cancer. Exposure to rock or slag wool may increase the risk of such cancers; however, the data do not convincingly prove that this association is causal."

Recent studies, including case-control studies, make it clear that any lung-cancer SMRs based on national data must take into account the potential confounding effect of smoking. Evidence from the case-control studies demonstrates that there is no significant association between fiber exposure and lung cancer or nonmalignant respiratory disease in the MVF manufacturing environment. It is clear, for example, that of the Newark, Ohio, plant workers (who made up some 35% of the U.S. cohort) exposure to MVF, including respirable glass fibers, was not responsible for any increase in lung cancer risk (Chiazze et al. 1993).

A recent agreement between the Occupational Safety and Health Administration and manufacturers and users of MVF for monitoring of the exposure of production workers and users might provide important exposure data for future studies (NAIMA 1999). That would be particularly valuable in light of the sparse epidemiological data on special-purpose fibers and RCF. More data are also needed on exposure of women and members of racial and other ethnic minorities to MVF. Nested case-control studies that can provide information on potential confounders might be the best means of addressing continued questions surrounding the effects of exposures to MVF.

Both cohort and case-control methods have been used to describe the distribution of disease in a population and to derive statistical associations in the epidemiology of MVF. Regardless of the approach, to be useful, epidemiological investigations must demonstrate internal and external validity and freedom from bias. In the Navy review of epidemiological studies, insufficient consideration was given to either internal or external validity. Internal validity refers to the design and conduct of a study that is free of systematic errors or biases, which are always of concern in epidemiological investigations. A number of biases, such as exposure and diagnostic misclassification, and confounding by lack of data on cigarette smoking are limitations in many of the available epidemiological investigations, but little consideration is given to such potential problems in the Navy review of the epidemiological studies of MVF. For the Navy, external validity refers to the generalizability or applicability of results from the study populations to Navy personnel working with MVF. However, because Navy personnel are largely involved in the removal of MVF, their exposure could differ—in physicochemical properties of the fibers and in the health status of workers—from exposures experienced by manufacturing workers, who have been the focus of epidemiological investigations. Therefore, the difference in exposures between the two workforces raises concerns about inferring health risks to Navy personnel on the basis of data derived from the MVF manufacturing industry.

7

EVALUATION OF THE NAVY'S EXPOSURE STANDARD

N 1995, the Navy's rationale for adopting an exposure standard of 2 fibers/ cm^3 for manufactured vitreous fibers (MVF) stemmed from its desire to establish a standard that was as strict as or stricter than occupational exposure levels proposed or used by other organizations (Navy Environmental Health Center 1997b; Krevonick 1998). The exposure standard of 2 f/cm^3 was based on the systematic adoption or incorporation of existing laws and standards as defined in the Navy's 1994 *Occupational Safety and Health Program Manual* (CNO 1994). As stated in the manual, "...instructions based on these standards may simply refer to a specific Occupational Safety and Health Administration (OSHA) standard (e.g., 29CFR1910.95) or may paraphrase, transpose, or otherwise adopt the standard without altering the basic criteria" (CNO 1994).

When the Navy adopted as its standard for MVF the exposure level of 2 f/ cm^3 as measured by phase-contrast optical microscopy (PCOM), the National Institute for Occupational Safety and Health (NIOSH) had a recommended exposure limit (REL) of 3 f/cm^3, which had been established in 1977, whereas OSHA had proposed permissible exposure limit (PEL) of 1 f/cm^3 in 1992 (Krevonick 1998). The OSHA PEL was later withdrawn on the basis of the remanding of the entire air contaminants rule under which the PEL had been proposed. The NIOSH REL was based on exposure information from four epidemiological studies (Dement 1976; Hill et al. 1973; Konzen 1976; Fowler et al. 1971) that found an absence of health effects in workers exposed to glass fibers, at a mean concentration of about 3 f/cm^3 (NIOSH 1977). OSHA's proposed PEL was based on the risk of nonmalignant respiratory disease (OSHA 1992) and was supported by results of several epidemiological studies, includ ing Bayliss et al. (1976) and Enterline et al. (1987). OSHA was also concerned with potential carcinogenic effects of refractory ceramic fibers (RCF). In evaluating the existing exposure level, the Navy reviewed the scientific data on MVF but concluded that neither 3 f/cm^3 nor 1 f/cm^3 was necessarily scientifically justified. It selected a standard of 2 f/cm^3 because this it was more stringent than the NIOSH REL but not as stringent as the OSHA PEL (Krevonick 1998). The Navy selected an exposure level that was the average of the two existing occupational exposure limits.

In January 1999, the Navy revised its *Occupational Safety and Health Program Manual,* changing the occupational exposure limit for MVF to the American Conference of Governmental Industrial Hygienists (ACGIH) threshold limit value (TLV) of 1 f/cm^3 (CNO 1999). The Navy then requested that the subcommittee address several questions concerning the selection and adoption of the original 1995 standard of 2 f/cm^3 and the appropriateness of the newly established standard of 1 f/cm^3. These questions were: Are there any studies in the scientific literature that the Navy should have considered that would have justified an exposure standard of more than or less than 2 f/cm^3? Is the process that the Navy used to adopt the 2-f/cm^3 standard appropriate for MVFs? Does the subcommittee feel that the reduction in the exposure standard from 2 to 1 f/cm^3 is justified on the basis of current scientific evidence?

The subcommittee believes that the Navy did not fail to consider any specific scientific studies that would have justified selecting an exposure standard that was higher or lower than 2 f/cm^3 when it adopted this standard in 1995. However, the subcommittee does not consider the method used by the Navy to adopt the 2-f/cm^3 standard to be justifiable on the basis of the exposure limits given by OSHA and NIOSH, nor was it scientifically defensible. The Navy could have chosen to adopt one of the existing exposure limits (that is, NIOSH or OSHA), as stated in the Navy *Occupational Safety and Health Program Manual* (CNO 1994). If the Navy had elected to use an existing limit, it should have considered selecting the more conservative exposure limit of 1 f/cm^3. This exposure limit would be more protective of workers, particularly in light of the toxicological data that suggested the carcinogenic potential of RCF (Davis et al. 1983; Mast et al. 1995a; Mast et al. 1995b; McConnell et al. 1995). Alternatively, if the Navy intended to derive a new exposure standard, a risk assessment based on a rigorous scientific evaluation of the toxicological and epidemiological data should have been conducted to develop an exposure limit at which potentially no adverse health effects would be observed. Although the Navy's document reviews much of the relevant literature on physical, chemical, toxicological, and epidemiological data on MVF, the Navy did not conduct the kind of systematic analysis, integration, and thorough evaluation of alternative arguments that are necessary for the kind of scientifically based risk analysis within which the Navy hopes to couch its exposure standards. Nor was the Navy able to find the scientific justification that it sought for its proposed standard through reference to a rigorous risk analysis conducted and documented by another organization.

It must be acknowledged that developing an exposure standard for MVF based on a quantitative analysis of potential health risks is a great challenge, given the scientific issues and the incomplete and inconsistent data available. Nonetheless, analyses by Maxim et al. (1999a), Fayerweather et al. (1997), Moolgavkar et al. (1999), and Wilson et al. (1999) show that animal studies can be examined quantitatively, with appropriate caveats, to develop a low-dose exposure standard that aims to protect humans from chronic health effects, such as chronic obstructive pulmonary disease (COPD), fibrosis, and cancer. Elements of those analyses bear discussion, but they generally show that an exposure standard of around 1 or 2 f/cm^3 should result in relatively small, if any, chronic health risks. For example, Maxim et al. (1999a) found that exposure to two new MVF (Isofrax and Insulfrax) at 1 f/cm^3 would result in a working-lifetime cancer risk of less than 10^{-5}. Moolgavkar et al. (1999) reported that exposure to kaolin-based RCF at 1 f/cm^3 would be expected to result in an excess probability of lung cancer of 3.7×10^{-5} for workers at the age of 70 years. Wilson et al. (1999) analyzed risks of lung cancer in blown-glass workers and found that risks ranged from 2.0×10^{-6} for nonsmoking workers blowing glass fibers with a binder to 2.4×10^{-4} for smokers blowing glass fibers without a binder; exposures were assumed to be at 1 f/cm^3. The analyses also suggest that the lack of clear responses in the existing epidemiological studies is to be expected, given typical occupational exposure.

The Maxim et al. (1999a) and Fayerweather et al. (1997) analyses highlight some critical concerns:

- The lack of pronounced responses in rat and hamster inhalation bioassays is itself informative in limiting the tenable estimates of potency.

- It is possible to consider the implications of exposures to less durable fibers by examining risks posed by more durable fibers, as long as an allowance for biopersistence is made.
- Although critical assumptions must be made about dose-response curves and cross-species doses, the sensitivity of the outcome to the choice of risk assessment model can be illuminated with appropriate analyses.

Apart from those specific concerns, some key issues must be carefully examined:

- The biological mechanisms by which fibers might induce tumors must be examined for their impacts on the dose-response relationship. In particular, the relationship of particle overload and inflammation to high-dose effects observed in animals and the expected role of those processes at lower doses should be investigated. Existing models accommodate such processes poorly.
- The sensitivity of results to the selection of a deposition model must be considered, and the differential deposition patterns in rats and humans assessed.
- The considerable advances in understanding of fiber-clearance processes need to be incorporated into the analyses. These advances include better assessment of biopersistence, of the distribution of fiber dimensions over time, and of the impact of that distribution process on toxicity.
- Risk estimates are markedly influenced by the units that are used to express the dose of fibers when extrapolating from animals to humans (for example, fibers per unit surface area or fibers per unit volume of lung tissue).
- The presumption of tumorigenic equivalence across species for a specific achieved fiber concentration (after adjusting for biopersistence) needs to be examined, and potential alternatives need to be explored.
- Existing analyses focus on tumorigenicity, but fibrosis, bronchitis, and other potential chronic noncancer responses might be important in evaluating the health protectiveness of a fiber exposure standard.

In sum, even if the Navy were to choose to adopt an existing quantitative analysis, further discussions and exploration of assumptions would be in order. Accordingly, it is difficult for the subcommittee to provide a detailed review of the Navy's assessment and conclusions—in particu lar, to address explicitly the question regarding the scientific validity of the Navy's proposed standard for exposure to MVF. The Navy's documentation does not articulate any specific risk analysis, nor does it attempt to justify its choice of exposure standards on the basis of risk-based goals.

The subcommittee does, however, support the Navy's recent adoption of the 1-f/cm^3 standard in accordance with the Navy's *Occupational Safety and Health Program Manual* (CNO 1999). The ACGIH TLV is based on concern for irritation of the upper respiratory tract (Konzen 1980). Furthermore, the 1-f/cm^3 standard is also in accordance with OSHA's recommended PEL of 1 f/cm^3 for glass fibers and rock and slag wools. An exposure standard of 1 f/cm^3 has also been endorsed by the North American Insulation Manufacturers Association and the Building and Construction Trades Department of the American Federation of Labor-Congress of Industrial Organizations (ACGIH 1997), providing further support for this exposure limit. That endorsement, which stems from a formal agreement between manufacturers, fabricators, installers, and removers of MVF, is based on the feasibility of achieving the standard in the workplace by the manufacturers and users of MVFs, and not on an assessment of risk (NAIMA 1999). An exposure standard of 1 f/cm^3 is also in agreement with standards recently adopted by New Zealand, Norway, Sweden, and the province of Alberta, Canada (ACGIH 1997; see Table 7-1). The subcommittee notes, however, that none of the existing MVF exposure limits is based on a quantitative analysis of risks of any chronic respiratory disease, including cancer.

TABLE 7-1 Occupational Exposure Limits for MVF

Country	Limit (f/cm^3)
Australia	0.5
Canada (Alberta)	1.0 (0.5 for RCF)
Denmark	2.0
Germany	0.5
Netherlands	3.0 (1.0 for RCF)
New Zealand	1.0
Norway	1.0
Poland	2.0
Sweden	1.0

Having reviewed the Navy's document, the subcommittee finds that the coverage of the relevant literature is, in some cases, outdated and incomplete. In particular, recent advances in understanding the basis and course of fiber dissolution and clearance rates in the respiratory tract are not fully covered, and recent quantitative analyses from animal bioassays are not addressed. Strengths and limitations of the epidemiological data are not sufficiently explored. For example, the Navy does not consider the impact of known or likely biases in the cited studies, such as exposure and diagnostic misclassification, confounding caused by lack of data on smokers, and, most important, the differences in exposures between workers who manufacture MVF and Navy personnel and contractors who are engaged primarily in removing MVF. However, even if those shortcomings in the review of the literature were remedied, the integration and analysis of information needed to establish a sound, risk-based scientific basis for the proposed standard has yet to be undertaken. The subcommittee believes that if the review of the toxicological, epidemiological, and risk-assessment literature were updated, the resulting data set would support a more extensive risk analysis than has so far been attempted.

The subcommittee notes that although a 1-f/cm^3 standard is appropriate for MVF that include conventional glass fibers and rock and slag wools, it is unlikely to be sufficiently protective for MVF that are more biopersistent or have a high proportion of respirable fibers that are much smaller in diameter than the conventional glass fibers and rock and slag wools, which average at least 6 μm. For instance, some RCF and other durable MVF are sufficiently biopersistent to warrant a more restrictive exposure limit. Furthermore, some specialty fibers are both more biopersistent and smaller diameter than conventional glass fibers; many long glass fibers have average diameters less than about 0.3 μm and cannot be seen by PCOM analysis. Inhalation toxicity studies in rats and hamsters have demonstrated the carcinogenic effects of exposures to RCF and durable fiber glass that are currently regulated under the Navy's MVF standard (Gelzleichter et al. 1999; Mast et al. 1995a,b; McConnell et al. 1995, 1999).

The hazard associated with long, biopersistent fibers is related to the number concentration of all the long fibers, not just those measured by PCOM. For conventional glass fibers, rock and slag wools, and RCF, nearly all the fibers are resolvable by PCOM. For specialty glass fibers that are thin and biopersistent, the Navy should use scanning electron microscopy analyses that record the number concentrations of fibers longer than 5, 10, and 20 μm.

Thus, in addition to supporting the Navy's adoption of the 1-f/cm^3 standard for conventional glass fibers and rock and slag wools, the subcommittee believes that the Navy should consider establishing separate exposure standards for the more biopersistent MVF in recognition of their greater hazard potential. In 1997, ACGIH proposed an exposure level of 0.1 f/cm^3 for RCF, which was amended to a proposed 0.2 f/cm^3 in 2000; but ACGIH has not provided a scientific rationale for either of these levels. Furthermore, ACGIH has not moved this recommendation from the Notice of Intended Change to the adopted list. However, regulatory agencies in other countries have examined the issue and adopted standards for RCF that generally range from 0.5 to 1.0 f/cm^3 (Table 7-1). In 1997, the Refractory Ceramic Fiber Coalition (RCFC), a trade organization of U.S. RCF manufacturers, adopted a recommended exposure guideline of 0.5 f/cm^3 (RCFC 1997). Given the extensive global review of this issue, it would be reasonable for the Navy to adopt a more stringent exposure level for the more biopersistent MVF.

In May 1993, the RCFC and the U.S. Environmental Protection Agency (EPA) entered into a voluntary exposure-monitoring consent agreement. Under the 5-year agreement, RCFC obtained and analyzed over 4,500 personal occupational-exposure monitoring samples of RCF from manufacturing process through user installation and removal. Across the lifecycle categories of RCF, average ambient concentrations ranged from 0.1 to 1.1 f/cm^3. A proposed RCFC-EPA-OSHA-NIOSH agreement to continue the exposure-monitoring program is under discussion.

In light of the recently revised exposure standards for MVF that reflect improved emission-control measures and new scientific evidence of the adverse effects of MVF, particularly RCF, the subcommittee believes that the Navy should periodically reevaluate its exposure standard for MVF. The MVF exposure levels developed by OSHA, NIOSH, and ACGIH should be evaluated in an effort to maintain Navy exposure guidelines that are consistent with those adopted by these organizations.

One other issue that should be addressed by the Navy is the extent to which any standard developed by others on the basis of epidemiological studies of workers in MVF manufacturing is protective of Navy personnel engaged in the installation, maintenance, or removal of MVF in confined spaces. Navy personnel typically use specific glass fibers and wools for insulation purposes and continuous-glass fibers for reinforcement appli cations. For these workers, exposures can be highly variable in terms of concentration and relatively acute, in contrast with the generally continuous and longer-term exposures that take place in the manufacturing industry. For RCF that have experienced repeated thermal stress, the resulting airborne fibers can have altered physical and chemical properties, including length distributions and crystalline forms that affect their biopersistence (TIMA 1993, p. 36). Non-RCF MVF, however, melt when exposed to heat so transformation to crystalline forms is of less concern (TIMA 1993, p. 28). The Navy should evaluate how appropriate the exposure standards are for protecting the health of the its personnel who are exposed to MVF that might have been altered by thermal stress.

8

INFORMATION GAPS AND RECOMMENDATIONS FOR FUTURE RESEARCH

As is evident from the number of recent studies cited in this report, a considerable amount of research has been and is being conducted in North America and Europe to elucidate the health effects of exposure to manufactured vitreous fibers (MVF). Although much of the research might address the information gaps discussed below, the need for and direction of future research must nonetheless be emphasized.

A recent agreement between the Occupational Safety and Health Administration and manufacturers and users of MVF to monitor exposures of production workers and users, might provide important data for future epidemiological studies (NAIMA 1999). Results of recent life-cycle studies on refractory ceramic fibers (RCF) as used in consumer products (Venturin et al. 1997) are an excellent starting point for this type of study. Monitoring information will be particularly valuable in light of the sparse epidemiological data on special-purpose fibers and other MVF, especially fibers that have been thermally aged. Ideally, monitoring will yield much-needed data on exposures of women and members of racial and other ethnic minorities to MVF. Nested case-control studies that obtain information on potential confounders, such as smoking, could be the best means of addressing questions on the effects of exposures to MVF.

The extrapolation of risks of lung tumors in rats after chronic inhalation of fibers is based on the assumption that the sensitivity of rats to fiber-induced carcinogenesis is similar to that of humans. Potential differences in target-cell sensitivity between rats and humans and the impact of these differences on risk models need to be addressed in future research. The research should focus on elucidating the mechanisms of fiber carcinogenesis, including the reasons why hamsters are prone to developing pleural mesotheliomas but not lung tumors after chronic inhalation but rats are more likely to develop lung tumors and have a lower incidence of mesotheliomas. These studies should provide data to help determine the rodent species that are most appropriate for use in human-health risk assessment after exposures to fibers.

Bioassays for lung tumors after inhalation of fibers are based on expensive and time-consuming 2-year carcinogenicity studies. Although the intracavitary-exposure tests might cost less and produce a high incidence of mesotheliomas, they have substantial shortcomings, as discussed in Chapter 5. Consequently, the subcommittee notes that results of intracavitary exposure studies should not be used for purposes of risk assessment. Thus, there is a need for the development and validation of short-term assays to predict long-term effects. These predictive short-term tests could include inhalation and possibly in vitro studies once the mechanisms of fiber carcinogenicity are more fully understood. For any short-term test to be accepted, it must be properly validated against the results of chronic-inhalation studies.

Animal studies of the effects of chronic inhalation of MVF have used mainly particular animal species and fiber types. However, variation in sensitivity among species and dependence of response on fiber type and dimensions are evident. More systematic testing would clarify the patterns of species and fiber-type dependence and aid in the generalization of effects and application to human risk assessment. One such assessment has already been conducted to determine whether a glass-fiber insulation product required a cancer warning label under California proposition 65 (Fayerweather et al. 1997).

Dosimetry studies in animals are needed to clarify the nature and degree of deposition of fibers of various dimensions in different parts of the respiratory tract. Dosimetry studies should also examine the rates of translocation of deposited fibers and the rates at which fibers are dissolved or otherwise cleared; dissolution and cleaving change the magnitude and size distribution of local tissue burdens over time. Studies are needed to clarify how differences in those morphological processes result in variations in sensitivity among species. Methods for defining toxicologically equivalent exposures and tissue burdens across species are needed, and the application of these methods in extrapolating from animal results to humans should be assessed.

The biological mechanisms by which fibers induce tumors and nonma lignant pulmonary dysfunction need to be further investigated. In particular, factors that can influence the shape of the dose-response relationship and could help to identify exposures below which key biological responses are not to be expected, need to be identified. The effects of continuous versus intermittent exposures should also be examined.

In general, methods and data appropriate for pursuing quantitative analysis of potential risks of chronic disease from MVF inhalation are needed. Quantitative risk analysis based on epidemiological data is hampered by the limited quantitative information on dosimetry. Epidemiology studies that incorporate measurements of fiber exposures or that reconstruct estimates of worker exposures to fibers, over the last several decades would aid in this endeavor.

The use of MVF as insulation material means that they are subjected to variations in temperature. Heat stress can result in changes in fibers' chemical and physical structure. However, there is a lack of information on the effect of heat stress on fiber structure and fiber breakdown. Such information could be important for assessing the exposures of Navy personnel and other workers to MVF. The toxicity of heat-stressed MVF—particularly those produced to withstand high temperature, such as RCF—has yet to be fully explored.

Pursuit of a research agenda as suggested above poses formidable challenges. It will need to be a continuing process that is not the province solely of the Navy, but rather of the fiber research community in general. Although monitoring data might always be difficult to come by, it is imperative that the Navy determine the concentrations and characteristics (such as, fiber length, diameter, and composition) of the MVF, particularly RCF, to which its personnel are exposed. That could be accomplished by air monitoring or by an inventory of the fiber content of the MVF-containing equipment and materials used by the Navy.

The subcommittee believes that when the Navy adopts another organization's standard, as it has adopted the American Conference of Governmental Industrial Hygienists threshold limit value, it must evaluate the rationale used by the other organization to develop the standard so that it can assess how appropriate the standard is for its own purposes. It is important that typical fiber concentrations and exposure ranges be measured in the workplace air so that actual worker exposure can be compared with established or proposed occupational exposure standards. Such data and comparisons are necessary for any quantitative risk assess ment and its accompanying uncertainty analysis. The Navy might also want to consider asking its suppliers to provide, where possible, materials that use low-biopersistence MVF.

The Navy should regard its current occupational standard as only one point on an ever-changing mosaic of exposure standards that will require periodic adjustments as new toxicological data become available, new monitoring studies are conducted, and new MVF are introduced into the workplace.

REFERENCES

ACGIH (American Conference of Governmental Industrial Hygienists). 1997. Synthetic Vitreous Fibers. In: Supplements to the Sixth Edition Documentation of the Threshold Limit Value (TLV) and Biological Exposure Indices. American Conference of Governmental Industrial Hygienists, Cincinnati, OH.

Bader, M.E., R.A. Bader and I.J. Selikoff. 1961. Pulmonary function in asbestosis of the lung, an alveolar-capillary block syndrome. Amer. J. Med. 30(Feb.):235-242.

Barrett, J.C., P.W. Lamb and R.W. Wiseman. 1989. Multiple mechanisms for the carcinogenic effects of asbestos and other mineral fibers. Environ. Health Perspect. 81:81-89.

Bayliss, D.L., J.M. Dement, J.K. Wagoner, and H.P. Blejer. 1976. Mortality patterns among fibrous glass production workers. Ann. N Y Acad. Sci. 271:324-335.

Bellmann, B., H. Muhle, F. Pott, H. Konig, H. Kloppel, and K. Spurny. 1987. Persistence of man-made mineral fibres (MMMF) and asbestos in rat lungs. Ann. Occup. Hyg. 31(4B):693-709.

Bernstein, D.M., C. Morscheidt, H.-G. Grimm, P. Thevenaz, and U. Teichert. 1996. Evaluation of soluble fibers using the inhalation biopersistence model, a nine-fiber comparison. Inhalation Toxicol. 8(4):345-385.

Berry, G. 1999. Models for mesothelioma incidence following exposure to fibers in terms of timing and duration of exposure and the biopersistence of the fibers. Inhalation Toxicol. 11 (2):111-130.

Boffetta, P., A. Andersen, J. Hansen, J.H. Olsen, N. Plato, L. Teppo, P. Westerholm and R. Saracci. 1999. Cancer incidence among European man-made vitreous fiber production workers. Scand. J. Work Environ. Health 25 (3):222-226.

Boffetta, P., R. Saracci, A. Andersen, P.A. Bertazzi, J. Chang-Claude, J. Cherrie, G. Ferro, R. Frentzel-Beyme, J. Hansen, J. Olsen, N. Plato, L. Teppo, P. Westerholm, P.D. Winter and C. Zocchetti. 1997. Cancer mortality among man-made vitreous fiber production workers. Epidemiology 8(3):259-268.

Brody, A.R., L.H. Hill, and B. Warheit. 1985. Induction of early alveolar injury by inhaled asbestos and silica. Fed. Proc. 44(10):2596-2601.

Chiazze, L. Jr. and D.K. Watkins. 1994. Epidemiology studies of synthetic vitreous fibers: methods used and current studies. Regul. Toxicol. Pharmacol. 20(3 Pt 2):S58-S67.

Chiazze, L. Jr., D.K. Watkins, and C. Fryar. 1997. Historical cohort mortality study of a continuous filament fiberglass manufacturing plant. I. White men. J. Occup. Environ. Med. 39 (5):432-441.

Chiazze, L. Jr., D. Watkins, C. Fryar and J. Kozono. 1993. A case-control study of malignant and non-malignant respiratory disease among employees of a fiberglass manufacturing facility: II. Exposure assessment. Br. J. Ind. Med. 50(8):717-725.

Chiazze, L., D.K. Watkins, C. Fryar, W. Fayerweather, J.R. Bender, and M. Chiazze. 1999. Mortality from nephritis and nephrosis in the fiberglass manufacturing industry. Occup. Environ. Med. 56(3):164-166.

CNO (Chief of Naval Operations). 1994. Navy Occupational Safety and Health (NAVOSH) Program Manual. OPNAVST 5100.23D. Chief of Naval Operations, Washington, D.C.

CNO (Chief of Naval Operations). 1999. Navy Occupational Safety and Health (NAVOSH) Program Manual. OPNAVST 5100.23E. Chief of Naval Operations, Washington, D.C.

Collier, C.G., O. Kamstrup, K.J. Morris, R.A. Applin, and I.A. Vatter. 1997. Lung clearance of experimental man-made mineral fibres. Preliminary data on the effect of fiber length. Ann. Occup. Hyg. 41(Suppl.1):320-326.

Crapo, J.D., S.L. Young, E.K. Fram, K.E. Pinkerton, B.E. Barry, and R.O. Crapo. 1983. Morphometric characteristics of cells in the alveolar region of mammalian lungs. Am. Rev. Resp. Dis. 128(2 Pt 2):S42-S46.

Davis, J.M.G., J. Addison, R.E. Bolton, K. Donaldson, A.D. Jones and A. Wright. 1983. The pathologic effects of fibrous ceramic aluminum silicate glass to rats. Pp. 124 in: Biological Effects of Man-Made

Mineral Fibers, Report on a WHO/IARC meeting, Copenhagen, Denmark, 20-22 April 1982. Copenhagen, Denmark: WHO.

Davis, J.M., J. Addison, R.E. Bolton, K. Donaldson, A.D. Jones and T. Smith. 1986. The pathogenicity of long versus short fiber samples of amosite asbestos administered to rats by inhalation or intraperitoneal injections. Br. J. Exp. Pathol. 67(3):415-430.

Dement, J. 1976. Environmental aspects of fibrous glass production and utilization. Pp. 97-109 in: Occupational Exposure to Fibrous Glass -Proceedings of a Symposium. HEW Publication No. (NIOSH) 76-151. U.S. Department of Health, Education, and Welfare, Public Health Service, Center for Disease Control, National Institute for Occupational Safety and Health, Washington, D.C. April.

Doll, R. 1955. Mortality from lung cancer in asbestos workers. Br. J. Ind. Med. 12:81-86.

Eastes, W., and J.G. Hadley. 1995. Dissolution of fibers inhaled by rats. Inhalation Toxicol. 7 (2):179-196.

Eastes, W., and J.G. Hadley. 1996. A mathematical model of fiber carcinogenicity and fibrosis in inhalation and intraperitoneal experiments in rats. Inhalation Toxicol. 8(4):323-343.

Eastes, W., R.M. Potter, and J.G. Hadley. 2000. Estimating the in-vitro glass fiber dissolution rate from composition. Inhalation Toxicol. 12(4):269-280.

EC (European Communities). 1997. Commission Directive 97/69/EC. Official Journal of the European Communities, L343/19. December.

EC (European Commission Joint Research Centre). 1999. Methods for the Determination of the Hazardous Properties for Human Health of Man-Made Mineral Fibres (MMMF), D.M. Bernstein and J.M. Riego Sintes, eds. EUR 18748 EN. Ispra, Italy: European Commission Joint Research Centre.

ECETOC (European Centre for Ecotoxicological and Toxicology of Chemicals). 1996. Toxicology of Man-Made Organic Fibers. Report No. 69. Brussels: ECETOC April.

Enterline, P.E., G.M. Marsh, V. Henderson, and C. Callahan. 1987. Mortality update of a cohort of USA man-made mineral fiber workers. Ann. Occup. Hyg. 31(4B): 625-656.

Fayerweather, W.E., J.R. Bender, J.G. Hadley, and W. Eastes. 1997. Quantitative risk assessment for a glass fiber insulation product. Regul. Toxicol. Pharmacol. 25(2):103-120.

Fowler, D.P., J.L. Balzer, and W.C. Cooper. 1971. Exposure of insula tion workers to airborne fibrous glass. Amer. Ind. Hyg. Ass. J. 32(2):86-91.

Fubini, B., A.E. Aust, R.E. Bolton, P.J.A. Borm, J. Bruch, G. Ciapetti, K. Donaldson, Z. Elias, J. Gold, M.C. Jaurand, A.B. Kane, D. Lison, and H. Muhle. 1998. Non-animal tests for evaluating the toxicity of solid xenobiotics. ATLA 26(5):579-617.

Gelzleichter, T.R., E. Bermudez, J.B. Mangum, B.A. Wong, D.B. Janszen, O.R. Moss, and J.I. Everitt. 1999. Comparison of pulmonary and pleural responses of rats and hamsters to inhaled refractory ceramic fibers. Toxicol. Sci. 49(1):93-101.

Gilson, J.C. 1966. Health hazards of asbestos. Recent studies on its biological effects. Trans. Soc. Occup. Med. 16(3):62-74.

Hansen, E.F., F.V. Rasmussen, F. Hardt and O. Kamstrup. 1999. Lung-function and respiratory health of long-term fiber-exposed stonewool factory workers. Am. J. Respir. Crit. Care Med. 160(2):466-472.

Heaney, P.J. and J.A. Banfield. 1993. Structure and chemistry of silica, metal oxides, and phosphates. Pp. 185-234 in: Health Effects of Mineral Dusts, Reviews in Mineralogy, Vol. 28, G.D. Guthrie Jr. and B.T. Mossman, eds. Washington, D.C.: Mineralogical Society of America.

HEI (Health Effects Institute). 1991. Asbestos in Public and Commercial Buildings: A Literature Review and Synthesis of Current Knowledge. Cambridge, MA: Health Effects Institute.

Henderson, R.F., K.E. Driscoll, J.R. Harkema, R.C. Lindenschmidt, I.Y. Change, K.R. Maples, and E.B. Barr. 1995. A comparison of the inflammatory response of the lung to inhaled versus instilled particle in F344 rats. Fundam. Appl. Toxicol. 24(2):183-197.

Hesterberg, T.W., G. Chase, C. Axten, W.C. Miller, R.P. Musselman, O. Kamstrup, J. Hadley, C. Morscheidt, D.M. Bernstein, and P. Thevenaz. 1998a. Biopersistence of synthetic vitreous fibers and amosite asbestos in the rat lung following inhalation. Toxicol. Appl. Pharmcol. 151(2):262-275.

Hesterberg, T.W., G.A. Hart, J. Chevalier, W.C. Miller, R.D. Hamilton, J. Bauer, and P. Thevenaz. 1998b. The importance of fiber biopersistence and lung dose in determining the chronic inhalation effects of X607, RCF1, and chrysotile asbestos in rats. Toxicol. Appl. Pharmacol. 153(1):68-82.

Hesterberg, T.W., C. Axten, E.E. McConnell, G.A. Hart, G.R. Chase, W. Miller, J. Chevalier, J. Everitt, P. Thevenaz, and G. Oberdorster. 1999.

Studies on the inhalation toxicology of two fiber glasses and amosite asbestos in the Syrian golden hamster. Part 1. Results of a subchronic study and dose selection for a chronic study. Inhalation Toxicol. 11(9):747-784.

Hesterberg, T.W., W.C. Miller, E.E. McConnell, J. Chevalier, J. Hadley, D.M. Bernstein, P. Thevenaz, and R. Anderson. 1993. Chronic inhalation toxicity of size-separated glass fibers in Fischer 344 rats. Fundam. Appl. Toxicol. 20(4):464-476.

Hill, J.W., W.S. Whitehead, J.D. Cameron, S.A. Hedgecock. 1973. Glass fibers - Absence of pulmonary hazard in production workers. Br. J. Ind. Med. 30:174-79.

IPCS (International Programme on Chemical Safety). 1988. Man-made Mineral Fibers. Environmental Health Criteria 77. Geneva, Switzerland: World Health Organization.

Kamstrup, O., J.M. Davis, A. Ellehauge, and M. Guldberg. 1998. The biopersistence and pathogenicity of man-made vitreous fibres after short- and long-term inhalation. Ann. Occup. Hyg. 42(3):191-199.

Kane, A.B. 1996. Mechanisms of mineral fibre carcinogenesis. IARC Sci. Publ. 140:11-34.

Kennedy, G.L., Jr, and D.P. Kelly. 1993. Introduction to Fiber Toxicology. Pp. 15-42 in: Fiber Toxicology, D.B. Warheit, ed. San Diego: Academic Press.

Konzen, J.L. 1976. Results of environmental air-sampling studies conducted in Owens-Corning fiberglass manufacturing plants. Pp. 115-120 in: Occupational Exposure to Fibrous Glass, A Symposium. HEW Pub. No.(NIOSH) 76-151. U.S. Department of Health, Education and Welfare, CDC, NIOSH, Washington, D.C.

Konzen, J.L. 1980. Man-made vitreous fibers and health. Pp. 329-342 in: Proceedings of the National Workshop on Substitutes for Asbestos. EPA 560/3-80-001. U.S. Environmental Protection Agency, Washington, DC.

Krevonick, P. 1998. Navy Exposure Limit for Man-Made Vitreous Fibers: A Retrospective Look at the Decision History. Navy Environmental Health Center, Norfolk, VA. October.

Krombach, F., S. Munzing, A.M. Allmeling, J.T. Gerlach, J. Behr, and M. Dorger. 1997. Cell size of alveolar macrophages: An interspecies comparison. Environ. Health Perspect. 105 (5):1261-1263.

Lee, I.M., C.H. Hennekens, D. Trichopoulos, and J.E. Buring. 1995.

Man-made vitreous fibers and risk of respiratory system cancer: a review of the epidemiologic evidence. J. Occup. Environ. Med. 37(6):725-738.

Lehnert, B.E., Y.E. Valdez, R.J. Sebring, N.M. Lehnert, G.C. Saunders, and J.A. Steinkamp. 1990. Airway intra-luminal macrophages: evidence of origin and comparison to alveolar macrophages. Am. J. Respir. Cell Mol. Biol. 3(4):377-391.

Lemasters, G.K., J.E. Lockey, L.S. Levin, R.T. McKay, C.H. Rice, E.P. Horvath, D.M. Papes, J.W. Lu and D.J. Feldman. 1998. An industry-wide pulmonary study of men and women manufacturing refractory ceramic fibers. Am. J. Epidemiol. 148(9):910-919.

Lippmann, M. 1990a. Effects of fiber characteristics on lung deposition, retention and disease. Environ. Health Perspect. 88:311-317.

Lippmann, M. 1990b. Man-made mineral fibers (MMMF): Human exposures and health risk assessment. Toxicol. Ind. Health 6(2):225-246.

Lippmann, M. 1993. Biophysical factors affecting fiber toxicity. Pp. 259-303 in: Fiber Toxicology, D.B. Warheit, ed. San Diego: Academic Press.

Lippmann, M. 1994. Deposition and retention of fibres: Effects on incidence of lung cancer and mesothelioma. Occup. Environ. Med. 51(12):793-798.

Lockey, J., G. Lemasters, C. Rice, K. Hansen, L. Levin, R. Shipley, H. Spitz, and J. Wiot. 1996. Refractory ceramic fiber exposure and pleural plaques. Am. J. Respir. Crit. Care Med. 154 (5):1405-1410.

Lockey, J., L.S. Levin, G.K. Lemasters, R.T. McKay, C.H. Rice, K.R. Hansen, D.M. Papes, S. Simpson and M. Medvedovic. 1998. Longitudinal estimates of pulmonary function in refractory ceramic fiber manufacturing workers. Am. J. Respir. Crit. Care Med. 157(4 Pt1):1226-1233.

Lundborg, M., U. Johard, A. Johansson, A. Eklund, R. Falk, W. Kreyling and P. Camner. 1995. Phagolysosomal morphology and dissolution of cobalt oxide particles by human and rabbit alveolar macrophages. Exp. Lung Res. 21(1):51-66.

Marsh, G.M., P.E. Enterline, R.A. Stone, and V.L. Henderson. 1990. Mortality among a cohort of US man-made mineral fiber workers: 1985 follow-up. J. Occup. Med. 32(7):594-604.

Marsh, G., R. Stone, A. Youk, T. Smith, M. Quinn, V. Henderson, L. Schall, L. Wayne and K. Lee. 1996. Mortality among United States rock wool and slag wool workers: 1989 update. J. Occup. Health Safety 12(3):297-312.

Mast, R.W., T.W. Hesterberg, L.R. Glass, E.E. McConnell, R. Anderson, and D.M. Bernstein. 1994. Chronic inhalation and biopersistence of refractory ceramic fiber in rats and hamsters. Environ. Health Perspect. 102(Suppl 5):207-209.

Mast, R.W., E.E. McConnell, R. Anderson, J. Chevalier, P. Kotin, D.M. Bernstein, P. Thevenax, L.R. Glass, W.C. Miller and T.W. Hesterberg. 1995a. Studies of the chronic toxicology (inhalation) of four types of refractory ceramic fiber in male Fischer 344 rats. Inhalation Toxicol. 7(4):425-467.

Mast, R.W, E.E. McConnell, T.W. Hesterberg, J. Chevalier, P. Kotin, P. Thevenax, D.M. Bernstein, L.R. Glass, W.C. Miller and R. Anderson. 1995b. Multiple-dose chronic inhalation toxicity study of size-separated kaolin refractory ceramic fiber in male Fischer 344 rats. Inhalation Toxicol. 7(4):469-502.

Maxim, L.D., R.W. Mast, M.J. Utell, C.P. Yu, P.M. Boymel, B.K. Zoitos, J.E. Cason. 1999a. Hazard assessment and risk analysis of two new synthetic vitreous fibers. Regul. Toxicol. Pharmacol. 30(1):54-74.

Maxim, L.D., D. Venturin, J.N. Allshouse. 1999b. Respirable crystalline silica exposure associated with the installation and removal of RCF and conventional silica-containing refractories in industrial furnaces. Regul. Toxicol. Pharmacol. 29(1):44-63.

McClellan, R.O., F.J. Miller, T.W. Hesterberg, D.B. Warheit, W.B. Bunn, A.G. Kane, M. Lippmann, R.W. Mast, E.E. McConnell, and C. F. Reinhardt. 1992. Approaches to evaluating the toxicity and carcinogenicity of man-made fibers: Summary of a workshop held November 11-13, 1991, Durham, North Carolina. Regul. Toxicol. Pharmacol. 16 (3):321-364.

McConnell, E.E. 1995. Advantages and limits of In vivo screening tests. Ann. Occup. Hyg. 39 (5):727-735.

McConnell, E.E., C. Axten, T.W. Hesterberg, J. Chevalier, W.C. Miller, J. Everitt, G. Oberdorster, G.R. Chase, P. Thevenaz, and P. Kotin. 1999. Studies on the inhalation toxicology of two fiberglasses and amosite asbestos in the Syrian golden hamster. Part II. Results of chronic exposure. Inhalation Toxicol. 11(9):785-835.

McConnell, E.E., O. Kamstrup, R. Musselman, T.W. Hesterberg, J. Chevalier, W.C. Miiller, and P. Thevenaz. 1994. Chronic inhalation study of size-separated rock and slag wool insulation fibers in Fischer 344/N rats. Inhalation Toxicol. 6(6):571-614.

McConnell, E.E., R.W. Mast, T.W. Hesterberg, J. Chevalier, P. Kotin, D.M. Bernstein, P. Thevenaz, L.R. Glass and R. Anderson. 1995. Chronic inhalation toxicity of a kaolin based refractory ceramic fiber (RCF) in Syrian golden hamsters. Inhalation Toxicol. 7(4):503-532.

Mereweather, E.R.A, and C.W. Price. 1930. Report on Effects of Asbestos Dust on the Lungs and Dust Suspension in the Asbestos Industry. London: H.M.S.O.

Miller, F.J. 1999. Letter to Eugene McConnell on biological half times of synthetic vitreous fibers from Fred J. Miller, Director, Respiratory and Neurotoxicology Research Programs, Chemical Industry Institute of Toxicology, Dec.8, 1999.

Moolgavkar, S.H., E.G. Luebeck, J. Turin, and L. Hanna. 1999. Quantitative assessment of the risk of lung cancer associated with occupational exposure to refractory ceramic fibers. Risk Anal. 19(4):599-611.

NAIMA (North American Insulation Manufacturers Association). 1999. Letter to Adam Finkel, Director of Health Standards, Occupational Safety and Health Administration on voluntary health and safety partnership program for fiber glass, rock and slag wool fiber products from Kenneth D. Mentzer, Executive Vice President of the North American Insulation Manufacturers Association (NAIMA), May 18, 1999.

NEHC (Navy Environmental Health Center). 1997a. Health Hazard Information Summary. Man-Made Vitreous Fibers. Navy Environmental Health Center. Norfolk, VA. June.

NEHC (Navy Environmental Health Center). 1997b. Man-Made Vitreous Fibers. Technical Manual NEHC-TM6290.91-1 Rev. A. Navy Environmental Health Center. Norfolk, VA. October.

NIOSH (National Institute of Occupational Safety and Health). 1977. Criteria for a Recommended Standard: Occupational Exposure to Fibrous Glass. DHEW (NIOSH) Pub. No. 77-152. U.S. Department of Health, Education and Welfare, CDC, NIOSH, Washington, D.C.

NIOSH (National Institute of Occupational Safety and Health). 1994. NIOSH Manual of Analytical Methods, 4th Ed. NIOSH Pub. No. 94-113. Cincinnati, OH: U.S. Department of Health and Human Services, National Institute of Occupational Safety and Health.

OSHA (Occupational Safety and Health Administration). 1992. Fibrous

Glass Including Refractory Ceramic Fibers. Fed. Regist. 57(114): 26195-26202.

Pott, F. 1980. Animal experiments on biological effect of mineral fibres. Pp.261-272 in: Biological Effects of Mineral Fibres, Vol.1., IARC Scientific Publication No. 30, J.C. Wagner and W. Davis, eds. Lyon, France: International Agency for Research on Cancer.

Pott, F. and K.H. Friedrichs. 1972. Tumors of the rat after intra peritoneal injection of fibrous dust. [in German]. Naturwissenschaften 59(7):318.

Pott, F., M. Roller, U. Ziem, F.J. Reiffer, B. Bellmann, M. Rosenbruch, and F. Huth. 1989. Carcinogenicity studies on natural and man-made fibres with the intraperitoneal test in rats. Pp. 173-179 in: Non-Occupational Exposure to Mineral Fibres, IARC Scientific Pub. 90. J. Bignon, J. Peto, and R. Saracci, eds. Lyon, France: International Agency for Research on Cancer.

Potter, R.M. and S.M. Mattson. 1991. Glass fiber dissolution in a physiological saline solution. Glastechnische Berichte 64(1):16-28.

Raabe, O.G., M.A. Al-Bayati, S.V. Teague, and A. Rasolt. 1986. Regional Deposition of Inhaled Monodisperse Coarse and Fine Aerosol Particles in Small Laboratory animals. NTIS/ PB86-116555. California Univ., Davis. October.

RCFC (Refractory Ceramic Fiber Coalition). 1997. The Recommended Exposure Guideline (REG) for RCF Has Been Changed from 1.0 Fiber/Cubic Centimeter (f/cc) of Air to 0.5 f/cc. RCF Coalition, Washington, D.C. October 31,1997.

Rice, C.H., J.E. Lockey, G.K. Lemasters, L.L. Levin, P. Staley and K.R. Hansen. 1997. Estimation of historical and current employee exposure to refractory ceramic fibers during manufacturing and related operations. Appl. Occup. Environ. Hyg. 12(1):54-61.

Rogers, R.A., J.M. Antonini, H. Brismar, J. Lai, T.W. Hesterberg, E.H. Oldmixon, P. Thevenaz. and J.D. Brain. 1999. In situ microscopic analysis of asbestos and synthetic vitreous fibers retained in hamster lungs following inhalation. Environ. Health Perspect. 107(5):367-375.

Sali, D., P. Boffetta, A. Andersen, J.W. Cherrie, J.C. Claude, J. Hasen, J.H. Olsen, A.C. Pesatori, N. Plato, L. Teppo, P. Westerholm, P. Winter, and R. Saracci. 1999. Non-neoplastic mortality of European workers who produce man made vitreous fibres. Occup. Environ. Med. 56 (9):612-617.

Searl, A., D. Buchanan, R.T. Cullen, A.D. Jones, B.G. Miller and C.A. Soutar. 1999. Biopersistence and durability of nine mineral fibre types in rat lungs over 12 months. Ann. Occup. Hyg. 43 (3):143-153.

Sebring, R.J. and B.E. Lehnert. 1992. Morphometric comparisons of rat alveolar macrophages, pulmonary interstitial macrophages, and blood monocytes. Exp. Lung Res. 18(4):479-496.

Simonato, L., A.C. Fletcher, and J.W. Cherrie, A. Andersen, P. Bertazzi, N. Charnay, J. Claude, J. Dodgson, J. Esteve, et al. 1987. The International Agency for Research on Cancer historical cohort study of MMMF production workers in seven European countries - extension of the follow-up. Ann. Occup. Hyg. 31(4B):603-624.

Stanton, M.F. and C. Wrench. 1972. Mechanisms of mesothelioma induction with asbestos and fibrous glass. J. Natl. Cancer Inst. 48(3):797-821.

Stanton, M.F., M. Layard, A. Tegeris, E. Miller, M. May, E. Morgan and A. Smith. 1981. Relation of particle dimension to carcinogenicity in amphibole asbestoses and other fibrous minerals. J. Natl. Cancer Inst. 67(5):965-976.

Stone, K.C., R.R. Mercer, P. Gehr, B. Stockstill and J.D. Crapo. 1992. Allometric relationships of cell numbers and size in the mammalian lung. Am. J. Respir. Cell Mol. Biol. 6(2):235-243.

Sussman, R.G., B.S. Cohen,, and M. Lippmann. 1991a. Asbestos fiber deposition in a human tracheobronchial cast. I. Experimental. Inhalation Toxicol. 3(2):145-160.

Sussman, R.G., B.S. Cohen, and M. Lippmann. 1991b. Asbestos fiber deposition in a human tracheobronchial cast. II. Empirical model. Inhalation Toxicol. 3(2):161-179.

TIMA (Thermal Insulation Manufacturers Association). 1993. ManMade Vitreous Fibers: Nomenclature, Chemical and Physical Properties. Thermal Insulation Manufacturers Association, Alexandria, VA.

Trethowan, W.N., P.S. Burge, C.E. Rossiter, J.M. Harrington, I.A. Calvert. 1995. Study of the respiratory health of employees in seven European plants that manufacture ceramic fibers. Occup. Environ. Med. 52(2):97-104.

Unifrax Corporation. 1999. Material Safety Data Sheet from the Unifrax Corporation, Niagara Falls, NY.

Venturin, D., G. Deren, and M. Corn. 1997. Quantitative industrial hygiene product life cycle analysis applied to refractory ceramic fiber

(RCF) consumer product applications. Regul. Toxicol. Pharmacol. 25(3):232-239.

Wagner, J.C., C.A. Sleggs and P. Marchand. 1960. Diffuse pleural mesothelioma and asbestos exposure in the north western cape province. Brit. J. Industr. Med. 17: 260-271.

Warheit, D.B. 1994. A review of some biophysical factors and their potential roles in the development of fiber toxicity. Regul. Toxicol. Pharmacol. 20(3):S113-S120.

Warheit, D.B., and M.A. Hartsky. 1994. Influences of gender, species and strain differences in pulmonary toxicological assessments of inhaled particles and/or fibres. Pp. 253-265 in: Toxic and Carcinogenic Effects of Solid Particles in the Respiratory Tract. U. Mohr, D.L. Dungworth, J.L. Mauderly, and G. Oberdörster, eds. Washington, D.C.: ILSI Press.

Watkins, D.K., L. Chiazze, and C. Fryar. 1997. Historical cohort mortality study of a continuous filament fiberglass manufacturing plant. II Women and minorities. J. Occup. Environ. Med. 39(6):548-555.

WHO (World Health Organization). 1985. Reference Method for Measuring Airborne Man-made Mineral Fibres (MMVF). Environmental Health Report No.4. Copenhagen, Denmark: World Health Organization, Regional Office for Europe

WHO (World Health Organization). 1992. European Program for Occupational Health 1992. Validity of methods for assessing the carcinogenicity of man-made fibres. Executive Summary of a WHO Consultation, 19-20 May 1992. Copenhagen: Denmark: World Health Organization, Regional Office for Europe.

Wilson, R., A.M. Langer, and R.P. Nolan. 1999. A Risk Assessment for Exposure to Glass Wool. Regul. Toxicol. Pharmacol. 30:96-109.

Zoitos, B.K. and P.M. Boymel. 1999. Isofrax: A new soluble high-temperature fiber. The American Ceramic Society Bulletin 78(12):57-62.

Zoitos, B.K., A. De Meringo, E. Rouyer, S. Thelohan, J. Bauer, B. Law, P.M. Boymel, J.R. Olson, V.R. Christensen, M. Gulberg, A.R. Koenig, and M. Perander. 1997. In vitro measurement of fiber dissolution rate relevant to biopersistence at neutral pH: An interlaboratory round robin. Inhalation Toxicol. 9:525-540.